PAKs, RAC/CDC42 (p21)-activated Kinases

PAKs, RAC/CDC42 (p21)-activated Kinases

Towards the Cure of Cancer and Other PAK-dependent Diseases

Edited by

Hiroshi Maruta,

NF/TSC Cure Org,
Melbourne,
Australia

ELSEVIER AMSTERDAM • BOSTON • HEIDELBERG • LONDON • NEW YORK • OXFORD
PARIS • SAN DIEGO • SAN FRANCISCO • SINGAPORE • SYDNEY • TOKYO

Elsevier
32 Jamestown Road London NW1 7BY
225 Wyman Street, Waltham, MA 02451, USA

First edition 2013

Notices
Knowledge and best practice in this field are constantly changing. As new research and
experience broaden our understanding, changes in research methods, professional
practices, or medical treatment may become necessary.

Practitioners and researchers must always rely on their own experience and knowledge in
evaluating and using any information, methods, compounds, or experiments described herein.
In using such information or methods they should be mindful of their own safety and the
safety of others, including parties for whom they have a professional responsibility.

To the fullest extent of the law, neither the Publisher nor the authors, contributors, or
editors, assume any liability for any injury and/or damage to persons or property as a
matter of products liability, negligence or otherwise, or from any use or operation of any
methods, products, instructions, or ideas contained in the material herein.

British Library Cataloguing-in-Publication Data
A catalogue record for this book is available from the British Library

Library of Congress Cataloging-in-Publication Data
A catalog record for this book is available from the Library of Congress

ISBN: 978-0-12-407198-8

For information on all Elsevier publications
visit our website at elsevierdirect.com

This book has been manufactured using Print On Demand technology. Each copy is
produced to order and is limited to black ink. The online version of this book will show color
figures where appropriate.

Working together to grow
libraries in developing countries

www.elsevier.com | www.bookaid.org | www.sabre.org

ELSEVIER   BOOK AID
International   Sabre Foundation

Contents

List of Contributors

Greg M. Cole Department of Neurology, University of California Los Angeles and Geriatric Research and Clinical Center, Greater Los Angeles Veterans Affairs Healthcare System, West Los Angeles Medical Center, Los Angeles, CA, USA

Graham Cote Queen's University, Kingston, ON, Canada

Sally A. Frautschy Department of Neurology, University of California Los Angeles and Geriatric Research and Clinical Center, Greater Los Angeles Veterans Affairs Healthcare System, West Los Angeles Medical Center, Los Angeles, CA, USA

Hong He University of Melbourne, Austin Health, Melbourne, Australia

Ramesh K. Jha Bioscience Division, Los Alamos National Laboratory, Los Alamos, NM, USA

Stefan Knapp Nuffield Department of Clinical Medicine, University of Oxford, Old Road Campus Research Building, Structural Genomics Consortium, Oxford, UK

Qiu-Lan Ma Department of Neurology, University of California Los Angeles and Geriatric Research and Clinical Center, Greater Los Angeles Veterans Affairs Healthcare System, West Los Angeles Medical Center, Los Angeles, CA, USA

Hiroshi Maruta NF/TSC Cure Organisation, Melbourne, Australia

Shanta M. Messerli Marine Biological Laboratory, Woods Hole, MA, USA

Masato Okada Research Institute for Microbial Diseases, Osaka University, Suita, Osaka, Japan

Sumino Yanase Daito Bunka University, Saitama, Japan

Fusheng Yang Department of Neurology, University of California Los Angeles and Geriatric Research and Clinical Center, Greater Los Angeles Veterans Affairs Healthcare System, West Los Angeles Medical Center, Los Angeles, CA, USA

Introduction: Pushing the Boundary Toward Clinical Application: After 35 Years, PAK Research Comes of Age

Hiroshi Maruta

NF/TSC Cure Organisation, Melbourne, Australia

PAK is a family of Ser/Thr kinases that are activated by RAS-related G proteins of 21 kDa (p21) called RAC and CDC42. Although the first mammalian PAKs (PAK1 and PAK2) were cloned by Ed Manser's group [1] in Singapore (Figure 1) around 1994, the first member of the PAK family was isolated in 1977 by our team at NIH in a soil amoeba as *Acanthamoeba* myosin I heavy chain kinase (MIHCK) [2], far before a series of small G proteins (p21) such as RAS and RAC/CDC42 were discovered during the 1980s. Myosin I is a small unconventional "single-headed" myosin that, unlike conventional double-headed myosins (myosin II), lacks the C-terminal tail and requires phosphorylation of its heavy chain by MIHCK for actin activation of its intrinsic ATPase activity [2]. Once myosin I is phosphorylated, its interaction with actin filaments (F-actin) triggers a rapid ATP hydrolysis, the acto-myosin complex microfilaments associated with the leading edge of amoebas contract, and so-called amoeboid movement or membrane ruffling occurs.

However, until mammalian PAKs were cloned, what triggered the activation of this kinase remained unknown. Since RAC and CDC42, the activators of mammalian PAKs, are activated by oncogenic RAS through PI-3 kinase and a few other signal transducers [3], it is most likely that both mammalian and amoebal PAKs are activated by RAS or its upstream activators, such as growth factors, chemoattractants, and their receptors on the cell surface. Like amoebal PAKs, mammalian PAKs phosphorylate the regulatory light chain of smooth-muscle myosin II and trigger actin activation of myosin II ATPase, leading to a rapid contraction of smooth muscles. As anticipated, it has been well established that oncogenic RAS causes hypertension through contraction of blood vessels' smooth muscle.

Shortly after the oncogenic RAS-RAC/CDC42-PAKs signaling pathway was basically established in the mid-1990s, Jeff Field's group [4,5] in Philadelphia confirmed that RAS indeed activates PAK1, and overexpression of the dominant negative (DN) mutant of PAK1 in both RAS-transformed fibroblasts and *NF*1-deficient malignant peripheral nerve sheath tumors, in which RAS is abnormally activated, can reverse their malignant phenotype both *in vitro* and *in vivo*, strongly suggesting that PAK1

Figure 1 Ed Manser's team cloning the first mammalian PAK genes.

is essential for RAS-induced malignant transformation (anchorage-independent growth) of cells. Although a gene therapy with DN PAK1 itself cannot be used for clinical applications, Field et al.'s historical 1997–1998 findings triggered the development of more effective PAK1 blockers (anti-PAK1 compounds) for the future therapy of RAS cancers such as pancreatic and colon cancers, which represent more than 30% of all human cancers [6], as well as neurofibromatosis (NF), which causes tumors on the skin, along the spinal cord, or in the brain [3]. Interestingly, an *NF2* gene product called merlin, whose dysfunction causes type 2 NF (NF2), was found to be the direct inhibitor of PAK1, and the growth of *NF2*-deficient tumors such as meningioma, schwannoma, and mesothelioma also requires PAK1 [3].

Shortly after Ed Manser's group [7] found in 1998 that an SH3 adaptor protein called PIX is essential for the activation of PAK1 in cells, and microinjection of PAK18, a Pro-rich motif of 18 amino acids (residues 186–203) in PAK1 that binds the SH3 domain of PIX, blocks both PAK1-PIX interaction and RAS/RAC-induced membrane ruffling [8], my own team in Melbourne (Figure 2) generated the first cell-permeable PAK1-specific inhibitor, called WR-PAK18, by linking the peptide vector WR of 16 amino acids and PAK18 to selectively block oncogenic PAK1 signaling in cells [9]. WR-PAK18 selectively inhibits malignant (anchorage-independent) growth of RAS-transformed cells, without any effect on normal (anchorage-dependent) cell growth, clearly indicating that PAK1 is not essential for the growth of normal cells [9]. This suggests that unlike many conventional anticancer chemotherapies such as cisplatin and 5-FU, DNA poisons that kill fast-growing cells, including normal cells such as hair and bone marrow cells, PAK1 blockers would not cause the side effect of killing normal fast-growing cells.

Around the turn of the twenty-first century, the first signal therapeutic, STI-571 (also known as "Gleevec"), developed by Novartis, was approved by the FDA for

Figure 2 My team at LICR developing a series of PAK1 blockers.

the treatment of two rare cancers: chronic myelogenous leukemia and gastrointestinal stromal tumor. These cancers' growth depends on the Tyr kinases ABL and c-Kit, respectively, which are selectively inhibited by Gleevec [10]. Gleevec turned out to be a miracle drug for these rare cancers, which represent only less than 0.1% of all human cancers, but this drug is not effective for RAS cancers and most of the remaining cancers. Thus, we and others have launched studies to develop or identify potent PAK1 blockers, mainly for the treatment of RAS cancers and NF tumors.

The first synthetic PAK1-specific inhibitor turned out to be CEP-1347, which was developed originally for Parkinson's disease (PD) by Kyowa Hakko and Cephalon through a specific chemical modification of the antibiotic K252a, a no-specific kinase inhibitor, because CEP-1347 somehow blocks the kinase JNK, which is essential for apoptosis of neuronal cells. However, this drug does not inhibit JNK directly, and we found that it directly inhibits two Rac/CDC42-activated kinase families, PAK and MLK, which are essential for JNK activation [11]. This drug blocks the growth of RAS-transformed and *NF*2-deficient tumor cells as well as breast cancer cells with the IC_{50} around 1 μM but has no effect on normal cell growth [3,11,12], indicating that breast cancers also require PAK1 for their growth. Independently, Rakesh Kumar's group [13,14] at the M.D. Anderson Cancer Center also confirmed that PAK1 is essential for the growth of breast cancers, which represent around 30% of all female cancers. However, RAS mutation is very rare in breast cancers. In other words, more than 60% of female cancers, including breast cancers and RAS cancers, are PAK1-dependent for their growth.

During the last decade, a series of PAK1 blockers were developed or identified, including synthetic compounds as well as a variety of natural products (antibiotics). Among them the natural ring peptide FK228 is the most potent PAK1 blocker. Developed by Fujisawa Pharmaceuticals in Japan, it is a potent HDAC (histone deacetylase) inhibitor [15] that eventually blocks both upstream and downstream action of PAK1 with the IC_{50} below 1 nM [12]. However, since FK228 is unable to pass the blood–brain barrier (BBB), it may be ineffective for brain tumors such as gliomas and tumors associated with NF and tuberous sclerosis (TSC), whose growth also requires PAK1.

Quite recently a potent PAK inhibitor called PF-3758309 (UnPAK309 in short) was developed by Pfizer Oncology [16]. This drug directly inhibits both PAK1 and PAK4–6 with the IC_{50} around 10 nM, and suppresses the growth of several PAK1-dependent cancers such as breast, colon and lung cancers, as well as melanoma, *in vivo*, with a daily dose of 10–20 mg/kg [16]. Thus, in a decade or so UnPAK309 and FK228 could become available for patients who suffer from a variety of PAK1-dependent solid tumors, which represent more than 70% of all human cancers, as well as NF and TSC tumors.

Rather surprisingly, several non-tumor diseases such as AIDS, malaria, flu, Alzheimer's (AD), Huntington's (HD), inflammatory diseases such as asthma and arthritis, hypertension, epilepsy, depression, schizophrenia and autism associated with fragile X syndrome (FXS) also turned out to be PAK1-dependent. Thus, the potential market value of these PAK1 blockers could be huge in the future.

One more surprise has been revealed recently. More than two dozen natural PAK1-blocking compounds such as caffeic acid phenethyl ester (CAPE) and curcumin, the major anticancer ingredients in propolis and Indian curry, respectively, turned out to activate the tumor-suppressive kinase AMPK. This kinase is rapidly activated through another tumor-suppressive kinase called LKB1 as soon as the cellular glucose level drops upon either fasting or calorie restriction (CR). Interestingly, Makoto Taketo's group [17] at Kyoto University recently found that LKB1 inactivates PAK1 and activates AMPK. Since AMPK is responsible for glucose uptake into cells and longevity [18,19], just like the CR, these PAK1 blockers/AMPK activators would be potentially useful for the treatment of type 2 diabetes and improvement of quality of life, eventually leading to a substantial extension of our healthy lifespan.

How can we systematically screen for such potent PAK1 blockers/AMPK activators in the short term? We have recently devised a rapid and inexpensive *in vivo* screening system using a strain of the transparent nematode *C. elegans* (CL2070) that rapidly expresses GFP when the HSP16 gene promoter is activated through the transcription factor FOXO, which is activated by AMPK or inactivated by PAK1 shortly after heat shock treatment [20].

However, this nematode screening system cannot exclude inhibitors of another oncogenic kinase called AKT, which can also activate HSP16 gene through FOXO. In my opinion, AKT inhibitors in general are not so suitable for cancer therapy, in particular the lifelong treatment of NF or TSC, because unlike PAK1, AKTs are known to be essential for the growth or survival of normal cells, as well as for glucose uptake from the bloodstream by a variety of cells [21]. Furthermore, deletion of the major AKT isoform (AKT-1) in mice causes a variety of cardiovascular defects, in particular severe heart failure, leading to the death of 50% of offspring in the first four days after their birth [22]. In other words, AKT inhibitors could cause a variety of side effects in human patients, including cardiovascular dysfunction, immunosuppression and type 2 (insulin-resistant) diabetes, Thus, any potent compounds selected by this quick nematode screening system should be tested in mammalian cell culture to clarify if they block AKT or PAK1, and hopefully to get rid of AKT inhibitors.

Unlike PAK4-null mice, which are embryonically lethal [23], PAK5-null mice develop normally and are fertile [24]. Similarly, PAK1-null mice as well as PAK1-null nematodes develop normally and are fertile [20,25]. More interestingly, mast cells derived from PAK1-null mice no longer respond to the lipopolysaccharide challenge, which normally causes degranulation, leading to a rapid discharge of calcium ions [24]. This finding suggests that PAK1 is essential for a variety of inflammatory diseases such as asthma and arthritis. In fact, FK228 (2.5 mg/kg), which eventually blocks PAK1, suppresses autoantibody-induced arthritis in mice [26]. More surprisingly, the PAK1-null strain (RB689) of nematodes lives significantly longer than its control counterparts (N2), as discussed in Chapter 7.

Do we need PAK1 for healthy life? Nobody will know until we find a living PAK1-null person. However, I have a feeling that at least after birth, PAK1 is no longer essential, and could serve even as a sort of ticking time bomb that, when activated, shortens our lifespan by causing a variety of diseases or disorders, including deadly cancers, perhaps ensuring that we will not live forever.

An additional surprise emerged recently on the role of PAK2. Thomas Wieland's group [27] at the University of Heidelberg reported that PAK2 is essential for catecholamine-induced cardiac hypertrophy. It was previously reported that PAK1-deficient mice tend to suffer from cardiac hypertrophy [28]. However, the contribution of PAK1 deficiency appears to be rather marginal compared with the hyperactivation of PAK2 in developing hypertrophy, as several natural PAK blockers such as curcumin and CAPE effectively reverse cardiac hypertrophy in mice or rats.

Interestingly, in 2004, Susumu Tonegawa's group [29] at MIT found that the simultaneous inhibition of all three members of the group 1 PAK family (PAK1–3) in the neonatal forebrain of mice by the autoinhibitory domain of PAKs significantly impaired their hippocampus-dependent long-term memory (LTM), clearly indicating that at least one of these three PAKs is essential for maintaining LTM. We heard recently that Tonegawa's colleagues at Afraxis in San Diego, in collaboration with the Russian CRO in Moscow, have developed a few potent PAK1-specific inhibitors (IC$_{50}$ around 10 nM) that pass the BBB, so that they can test their effect on LTM in mice as well as whether these PAK1-specific inhibitors can cure or delay a variety of the known PAK1-dependent brain diseases or disorders.

http://www.ctf.org/pdf/ddi/2010-07-002.pdf.

Afraxis CEO Outlines Biotech's Success With Ultra-Lean Pre-Clinical R&D in Russia

Celebrating such an exciting PAK1-specific inhibitor development, in the following eight chapters, world experts in their fields discuss in detail, and with deep insight, how PAKs, in particular the oncogenic kinases PAK1 and PAK4 and their blockers, may control our life and health in a variety of aspects, and how mammalian PAKs have functionally evolved from their ancestral origin(s) in unicellular organisms such as yeast and amoebas through a series of mutations over millions of years.

We dedicate this book to the late Gary Bokoch (1954–2010) at Scripps, our dearest old friend, who led this PAK field with his deep insight and great kindness.

Acknowledgments

The editor is very grateful to all contributors to this book, in particular Drs. Stefan Knapp of Oxford University and Qiulan Ma of UCLA, who provided us with their own chapters.

References

[1]   Manser E, Leung T, Salihuddin H, Zhao ZS, et al. A brain Ser/Thr protein kinase activated by Cdc42 and Rac1. Nature 1994;367:40–6.

[2]   Maruta H, Korn E. Acanthamoeba cofactor protein is a heavy chain kinase required for actin activation of the Mg^{2+}-ATPase activity of *Acanthamoeba* myosin I. J Biol Chem 1977;252:8329–32.

[3]   Hirokawa Y, Tikoo A, Huynh J, Utermark T, et al. A clue to the therapy of neurofibromatosis type 2: NF2/merlin is a PAK1 inhibitor. Cancer J 2004;10:20–6.

[4]   Tang Y, Chen Z, Ambrose D, Liu J, et al. Kinase-deficient Pak1 mutants inhibit Ras transformation of Rat-1 fibroblasts. Mol Cell Biol 1997;17:4454–64.

[5]   Tang Y, Marwaha S, Rutkowski J, Tennekoon G, et al. A role for Pak protein kinases in Schwann cell transformation. Proc Natl Acad Sci USA 1998;95:5139–44.

[6]   Maruta H, Burgess AW. Regulation of the Ras signalling network. Bioessays 1994;16:489–96.

[7]   Manser E, Loo T, Koh C, Zhao Z, et al. PAK kinases are directly coupled to the PIX family of nucleotide exchange factors. Mol Cell 1998;1:183–92.

[8]   Obermeier A, Ahmed S, Manser E, Ye SC, et al. PAK promotes morphological changes by acting upstream of Rac. EMBO J 1998;17:4328–39.

[9]   He H, Hirokawa Y, Manser E, Lim L, et al. Signal therapy for RAS induced cancers in combination of AG 879 and PP1, specific inhibitors for ErbB2 and Src family kinases, that block PAK activation. Cancer J 2001;7:191–202.

[10] Jones RL, Judson IR. The development and application of imatinib (Gleevec). Expert Opin Drug Saf 2005;4:183–91.

[11] Nheu T, He H, Hirokawa Y, Tamaki K, et al. The K252a derivatives, inhibitors for the PAK/MLK kinase family selectively block the growth of RAS transformants. Cancer J 2002;8:328–36.

[12] Hirokawa Y, Arnold M, Nakajima H, Zalcberg J, et al. Signal therapy of breast cancer xenograft in mice by the HDAC inhibitor FK228 that blocks the activation of PAK1 and abrogates the tamoxifen-resistance. Cancer Biol Ther 2005;4:956–60.

[13] Balasenthil S, Barnes C, Rayala S, Kumar R. Estrogen receptor activation at Ser 305 is sufficient to upregulate cyclin D1 in breast cancer cells. FEBS Lett 2004;567:243–7.

[14] Wang RA, Zhang H, Balasenthil S, Medina D, et al. PAK1 hyperactivation is sufficient for mammary gland tumor formation. Oncogene 2006;25:2931–6.

[15] Nakajima H, Kim YB, Terano H, Yoshida M, et al. FK228, a potent anti-tumor antibiotic, is a novel histone deacetylase inhibitor. Exp Cell Res 1998;241:126–33.

[16] Murray B, Guo CX, Piraino J, Westwick J, et al. Small-molecule PAK inhibitor PF-3758309 is a potent inhibitor of oncogenic signaling and tumor growth. Proc Natl Acad Sci USA 2010;107:9446–51.

[17] Deguchi A, Miyashi H, Kojima Y, Okawa K, et al. LKB1 inactivates PAK1 by phosphorylation of Thr 109 in the p21-binding domain. J Biol Chem 2010;285:18283–18290.

[18] Li HB, Ge YK, Zheng XX, Zhang L. Salidroside stimulated glucose uptake in skeletal muscle cells by activating AMPK. Eur J Pharmacol 2008;588:165–9.

[19] Greer E, Dowlatshahi D, Banko M, Villen J, et al. An AMPK-FOXO pathway mediates longevity induced by a novel method of dietary restriction in *C. elegans*. Curr Biol 2007;17:1646–56.

[20] Maruta H. An innovated approach to *in vivo* screening for the major anti-cancer drugs Horizons in cancer research, 41. : Nova Science Publishers; 2010. 249–59

[21] Kohn A, Summers S, Birnbaum M, Roth R. Expression of a constitutively active AKT in 3T3-L1 adipocytes stimulates glucose uptake and GLUT-4 translocation. J Biol Chem 1996;271:31372–31378.

[22] Chang Z, Zhang Q, Feng QT, Xu J, et al. Deletion of *Akt1* causes heart defects and abnormal cardiomyocyte proliferation. Dev Biol 2010;347:384–91.

[23] Qu J, Li X, Novitch B, Zheng Y, et al. PAK4 kinase is essential for embryonic viability and for proper neuronal development. Mol Cell Biol 2003;23:7122–33.

[24] Li X, Minden A. Targeted disruption of the gene for the PAK5 kinase in mice. Mol Cell Biol 2003;23:7134–42.

[25] Allen J, Jaffer Z, Park S, Burgin S, et al. PAK1 regulates mast cell degranulation via effects on calcium mobilization and cytoskeletal dynamics. Blood 2009;113:2695–705.

[26] Nishida K, Komiyama T, Miyazawa S, Shen ZN, et al. Histone deacetylase inhibitor suppression of autoantibody-mediated arthritis in mice via regulation of p16INK4a and p21WAF1/Cip1 expression. Arthritis Rheum 2004;50:3365–76.

[27] Vettel C, Wittig K, Vogt A, Wuertz C, El-Armouche A, Lutz S, et al. A novel player in cellular hypertrophy: G(i)βγ/PI3K-dependent activation of the RacGEF TIAM-1 is required for α(1)-adrenoceptor induced hypertrophy in neonatal rat cardiomyocytes. J Mol Cell Cardiol 2012;53:165–75.

[28] Liu W, Zi M, Naumann R, Ulm S, Jin J, Taglieri DM, et al. Pak1 is a novel signaling regulator attenuating cardiac hypertrophy in mice. Circulation 2011;124:2702–15.

[29] Hayashi ML, Choi SY, Rao BS, Jung HY, et al. Altered cortical synaptic morphology and impaired memory consolidation in forebrain-specific dominant-negative PAK transgenic mice. Neuron 2004;42:773–87.

1 Functional Maturation of PAKs from Unicellular to Multicellular Organisms

Masato Okada[1], Graham Cote[2], Ramesh K. Jha[3], and Hiroshi Maruta[4]

[1]Research Institute for Microbial Diseass, Osaka University, Suita, Osaka, Japan, [2]Queen's University, Kingston, ON, Canada, [3]Bioscience Division, Los Alamos National Laboratory, Los Alamos, NM, USA, [4]NF/TSC Cure Organisation, Melbourne, Australia

Abbreviations

AD	Alzheimer's disease
HD	Huntington's disease
LD	learning deficit
AID	autoinhibitory domain
CRIB	CDC42/RAC-interactive binding
GBD	GTPase-binding domain
MLK	mixed-lineage kinase
MIHCK	myosin I heavy-chain kinase
CSK	C-terminal SRC kinase
IR	insulin receptor
GIT	G protein-coupled receptor kinase interactor
Prp	prion protein
DN	dominant negative
CA	constitutively active
LTP	long-term potentiation

1.1 Introduction

Back in early January of 1994, in the year's first issue of *Nature*, volume 367 (No. 6458), pages 40–46, a team led by Ed Manser and Louis Lim at the National University of Singapore reported the cloning of two new mammalian genes (cDNAs) called p21-activated kinases (PAKs) that encoded Ser/Thr kinases. These first mammalian PAKs (PAK1 and PAK2) were found to be activated by two distinct small

PAKs, RAC/CDC42 (p21)-activated Kinases. DOI: http://dx.doi.org/10.1016/B978-0-12-407198-8.00001-1

GTPases (p21s) called RAC and CDC42 [1]. Interestingly, these PAKs are closely related to the budding yeast *S. cerevisiae* STE20 kinase, which had been cloned two years earlier by a group led by Ekkehard Leberer at National Research Council Canada [2]. STE20 is required to transmit the pheromone signal from the G βγ subunit of heterotrimeric G proteins to downstream signal transducers, including MAPK-family kinases. Overproduction of STE20 can suppress the mating defect of dominant negative (DN) G beta mutations. In other words, G βγ in yeast is functionally equivalent to RAC/CDC42 in mammals. STE20 can also be activated by the binding of Cdc42p to the CDC42/RAC-interactive binding (CRIB) domain, an interaction that is essential for filamentous growth [3]. Interestingly, SHK1 in the fission yeast *S. pombe* is closely related to STE20 and can be activated by CDC42 [4].

As these few examples have suggested, PAKs are ubiquitous kinases present in all eukaryotes from yeast to human [4]. There are two groups of PAKs in mammals: those in group 1 (PAK1–3) are activated by both RAC and CDC42, whereas those in group 2 (PAK4–6) are activated only by (or just bind) CDC42 [3]. There exists another p21 (RAC/CDC42)-activated kinase family in mammals called MLKs (mixed-lineage kinases), which consists of three members (MLK1–3) [5]. MLKs are functionally much closer to group 1 PAKs. Like PAK1–3, MLKs form an inactive homodimer through their N-terminal autoinhibitory domain (AID), and are responsible for the activation of c-Jun N-terminal kinase (JNK). Thus, PAK/MLK experts no longer use the original term "p21-activated kinases" for PAKs or MLKs, to avoid any confusion. In addition, there is another p21, an inhibitor of cyclin-dependent kinases (CDKs), which is completely unrelated to RAC and CDC42 [6,7].

In living cells, PAKs are regulated not only by RAC/CDC42, but also by several other intracellular signal transducers such as SH3 adaptor proteins (PIX and NCK), Tyr kinases (FYN and ETK), and tumor suppressors called merlin and nischarin that directly inhibit PAK1–2. The AID in the N-terminal half of PAK1–3 blocks the C-terminal kinase domain intermolecularly in *trans* by forming an antiparallel homodimer (Figure 1.1). According to the X-ray crystallographic study by Maria Parrini and her colleagues at Harvard [8], the PAK1 homodimer is catalytically inactive, and dissociated by CDC42/RAC for activation in the following mechanism. As shown in Figure 1.1, PAK1's CRIB motif near the N-terminal of the AID contains a motif of seven amino acids (residues 81–87) called Di, which is responsible for antiparallel homodimerization, taking place only in the absence of the GTPases. Furthermore, near the C-terminal of AID there is another motif (residues 136–149),

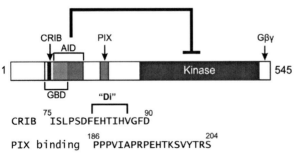

Figure 1.1 Human PAK1 domains.

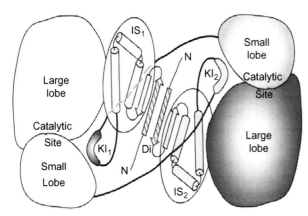

Figure 1.2 Autoinhibited homodimer of PAK1. With the permission of *Mol. Cell*, a part of figure 1B of Ref. [8] is reproduced here to show how the Di and Ki motifs of the AID are involved in maintaining the *trans* autoinhibited state of PAK1 in the antiparallel homodimer, which could be unfolded and activated upon GTPase binding.

called Ki, which is responsible for the *trans* kinase inhibition. On dimerization, the Ki of a PAK1 molecule binds the C-terminal kinase (catalytic) domain of its partner to mutually interlock (Figure 1.2). On the GTPase binding to the GBD (GTPase-binding domain), including the CRIB, this dimer is dissociated (unfolded), and each kinase domain becomes catalytically active. In fact, this is a rather simplified "test-tube" event, and in living cells, several other molecules are involved in both positive and negative regulation of PAK1–3, as briefly mentioned earlier in this chapter.

These highly sophisticated regulatory systems developed along with the functional evolution of PAK molecules *per se*, in particular via changes in their N-terminal half. In the following sections, we shall discuss how PAK molecules have functionally evolved (or matured), in particular from their first ancestor (origin) in unicellular organisms such as amoebas to their multicellular counterparts through the animal kingdom. No plants contain PAK, suggesting that PAKs play a key role in the regulation of actomyosin-based cell motility, which makes animal cells capable of moving around.

1.2 MIHCKs/PAKs and Myosin I in Unicellular *Acanthamoeba* and Multicellular *Dictyostelium*

The first PAK-family kinase was discovered in 1977 by our team at NIH as a myosin I heavy-chain kinase (MIHCK) from a freely living (unicellular) soil amoeba called *Acanthamoeba castellanii*[9,10]. Later MIHCK (now termed PAKB) was also found in *Dictyostelium discoideum* by another of our teams at Queen's University in Canada [11]. These amoebal MIHCK/PAKBs are essential for the unconventional single-headed class I myosins (Myosin I) to exhibit actin-activated Mg^{2+}-ATPase activity. They are functionally closer to the group 1 PAKs because they are activated by RAC/CDC42, and have an AID in the N-terminal half [12,13]. Like mammalian PAKs, they are able to phosphorylate the regulatory light chain to fully activate smooth muscle myosin II [14]. However, unlike mammalian PAK1–3, MIHCKs do not contain the Pro-rich motif that binds PIX, an SH3 adaptor protein [15]. Mammalian PAK1–3 must bind the PIX for full activation in cells. Thus, a Pro-rich peptide of 18 amino acids (PAK18) derived from the PIX-binding site serves as a specific inhibitor of PAK1–3 [16,17]. PAK18 does not, however, affect the kinase activity of either PAK4–6 or amoebal MIHCK/PAKB.

The first MIHCK was found during the so-called pre-oncogene era, and none of the small GTPases such as RAS, RHO, RAC, and CDC42 had yet emerged on the horizon. Thus, the major approach to the biological function of MIHCK was based on classic protein biochemistry *in vitro*. These studies revealed that MIHCK was activated by acidic phospholipids, and that the calcium–calmodulin complex antagonized lipid-mediated activation [18]. Later, when amoebal *MIHCK* genes were cloned, gene knockouts were used to identify their physiological function in the "social" amoeba *Dictyostelium*. This organism displays a unique life cycle depending on nutrient conditions. When nutrients (such as bacteria) are abundant, *Dictyostelium* grows as a freely living unicellular amoeba. However, shortly after they are starved, *Dictyostelium* amoebas aggregate to form a multicellular slug, navigated by pulsatile secretion of the chemoattractant cyclic AMP. Eventually each slug differentiates into two major cell types: fruiting body spores and stalks (Figure 1.3). *Dictyostelium* expresses four distinct PAKs (A–D). PAKB (MIHCK), as well as PAKA and PAKC, are expressed at both the unicellular vegetative and multicellular slug stages, but no information is yet available for PAKD.

The four *Dictyostelium* PAKs share very similar GBDs and kinase domains but otherwise are highly divergent. PAKB/MIHCK phosphorylates a critical site termed the TEDS rule site in the myosin I motor domain [19–21]. Myosins require a negative charge at the TEDS rule site to interact in a productive manner with F-actin (actin filament), which can be supplied by a Glu or Asp residue or by phosphorylation of a Ser or Thr residue. All seven *Dictyostelium* myosin I isozymes (MyoA–G and MyoK) have a Ser or Thr at the TEDS rule site and must be phosphorylated in order to efficiently move actin filaments *in*

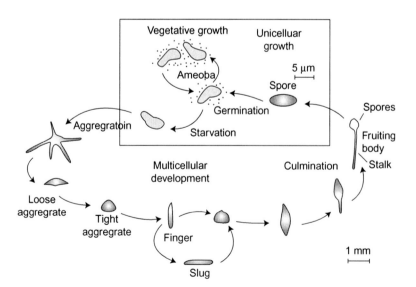

Figure 1.3 Life cycle of *Dictyostelium discoideum*.

vitro and to function *in vivo*[19]. PAKB can be activated by several different members of the *Dictyostelium* RAC/CDC42 family or by acidic phospholipids [22], and colocalizes with myosin I to the leading edge of migrating cells and to phagocytic cups. However, disruption of the *PAKB* gene results in only mild defects in chemotaxis and other myosin I-dependent processes [23]. Interestingly, PAKC-null cells extend multiple lateral pseudopodia and are unable to polarize properly in response to chemoattractants, a phenotype similar to cells that lack myosin I [24,25]. These chemotactic defects become much more severe in cells that lack both PAKB and PAKC, indicating that these two PAKs have at least partially overlapping functions [25].

The function of PAKA seems to be distinct from that of PAKB and PAKC. PAKA colocalizes with myosin II to the cleavage furrow of dividing cells and the posterior of polarized, chemotaxing cells via its N-terminal domain. PAKA is required for complete cytokinesis in suspension, to maintain the direction of cell movement to suppress lateral pseudopod extension, and to properly retract the posterior of the cell during chemotaxis. All of these defects can be traced to a loss of filamentous myosin II: the myosin II cap at the posterior of chemotaxing cells and myosin II assembly into cytoskeleton upon cAMP stimulation are absent in these cells. Moreover, constitutively active (CA) PAKA leads to an upregulation of myosin II assembly into filaments. In response to chemoattractant signaling, PAKA is transiently activated and is incorporated into the actin cytoskeleton with kinetics similar to those of myosin II assembly. Thus, PAKA appears to play a major role in promoting myosin II filament assembly. However, PAKA does not directly phosphorylate myosin II, and may function by inhibiting one of the *Dictyostelium* myosin II heavy-chain kinases, which block myosin bipolar filament formation [26,27].

As discussed in detail in the next section, *Dictyostelium* and *Acanthamoeba* PAKs may provide a unique example as to how these kinases could control the unicellular life style of *Acanthamoeba*, and the multicellular development of *Dictyostelium* by their distinct CRIB motif of 16 amino acids within the N-terminus of the GBDs. The CRIB motif binds RAC/CDC42 and eventually loosens the AID–kinase domain interaction, which blocks kinase activity (Figures 1.1 and 1.2). In general, when the CRIB motif binds tightly to CDC42/RAC, the *trans* interaction of the AID (through Di and Ki motifs) with the kinase domain is weakened, and the kinase domain is released from the AID-induced self-inhibited state, as discussed in the introduction of this chapter. This converts the inactive (closed antiparallel homodimer) form to the active (open) form, which is potentially oncogenic if expressed in normal mammalian fibroblasts.

1.3 PAK Evolution Among Metazoans (Unicellular Versus Multicellular)

1.3.1 *Choanoflagellates Versus Freshwater Sponges*

A number of mammalian oncogenes such as *SRC* and *RAS* are known to abolish the contact inhibition of normal cell growth during the malignant transformation of fibroblasts or epithelial cells by disrupting the gap junction through the GTPases RAC/CDC42 and the kinase PAK1 [28–30].

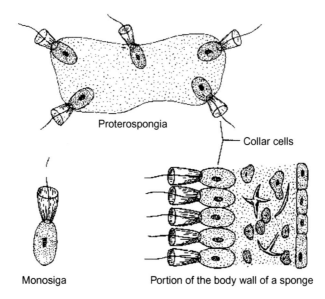

Figure 1.4 The unicellular choanoflagellate *M. ovata*. Morphological similarities between unicellular choanoflagellates and multicellular sponges are shown.

To understand further the role of these oncogenes in cell–cell interaction, in particular during multicellular tissue formation, our team at Osaka University has investigated the evolutionary changes in the self-regulatory mechanisms of two pro-oncogenic kinases, SRC and PAK, during the transition from unicellular ancestors such as the choanoflagellate *Monosiga ovata* (Figure 1.4) to primitive multicellular animals such as the freshwater sponge *Ephydatia fluviatilis* (Figure 1.5).

SRC and the C-terminal SRC kinase (CSK)-mediated regulatory system are established in the unicellular *M. ovata*, and we found that the unicellular SRC has unique features compared with those of its multicellular counterparts [31]. The former can be phosphorylated by CSK at the C-terminal self-regulatory Tyr residue, but still exhibits substantial activity even in the phosphorylated form. Analyses of chimeric molecules between *M. ovata* and *E. fluviatilis* SRC orthologs reveal that structural alterations in the kinase domain are responsible for the unstable negative regulation of *M. ovata* SRC [31]. When expressed in vertebrate fibroblasts, *M. ovata* SRC can induce cell transformation independent of CSK. These findings suggest that a structure of SRC required for stable CSK-mediated negative regulation still is immature in the unicellular *M. ovate*, and that the development of stable negative regulation of SRC may correlate with the evolution of multicellularity in animals.

Several other proto-oncogenic genes also regulate cellular homeostasis in metazoans and can be converted to oncogenes by gain-of-function mutations. To address the molecular basis for development of the PAK regulatory system during evolution, we cloned PAK orthologs from *M. ovata* (MoPAK) and from *E. fluviatilis* (EfPAK), and compared their regulatory features [32].

Figure 1.5 The multicellular freshwater sponge *E. fluviatilis*. Kindly provided by the photographer, Jiri Bohdal.

Like mammalian PAK1, EfPAK contains the GBD, AID, and Pro-rich PIX-binding motif in its N-terminal half (Figure 1.2). In contrast, MoPAK lacks the Pro-rich motif that binds PIX [32]. Moreover, MoPAK is constitutively active and induces a malignant transformation (anchorage-independent growth) when expressed in mammalian fibroblasts, whereas the PAKs (group 1) from multicellular animals (mammals and sponge) are strictly autoregulated and fail to transform mammalian fibroblasts. Further analysis of a series of chimeric PAK mutants between MoPAK and EfPAK revealed that an alteration in the CRIB motif of MoPAK confers higher constitutive kinase activity as well as greater affinity for GTPases (RAC/CDC42). This alteration in the CRIB motif is responsible for two malignant phenotypes, namely anchorage-independent growth and *in vivo* growth of xenograft in mice [32].

Thus, the evolutionally "immature" CRIB of MoPAK contributes to the disruption of multicellular tissue organization due to a loss of contact inhibition or gap junction.

Sequence comparisons of the CRIB motif (corresponding to residues 75–90 of human PAK1) within the GBD between multicellular and unicellular PAKs (Figure 1.6) shows that positions 8 (Glu 82) and 10 (Thr 84) (flanking the essential His 83 residue), in particular, are highly conserved among multicellular PAKs, whereas these two residues vary among the unicellular PAKs.

These results indicate that the maturation of the AID function (much higher affinity for the kinase domain) as well as the CRIB function (lower affinity for RAC/CDC42) are required for the development of the strict regulatory system of PAKs, and suggest a potential link between the establishment of the self-regulatory system of proto-oncogenes (*SRC* and *PAK*) and metazoan evolution.

1.3.2 Soil Amoeba Versus Cellular Slime Mold

In this context, it is of interest to examine the CRIB sequences of the four distinct *Dictyostelium* PAKs (A–D) because this cellular slime mold has both unicellular and multicellular stages, depending on nutrient conditions. Richard Firtel's group at the University of California, San Diego (UCSD) found that *Dictyostelium* PAKA is not absolutely essential for the starvation-triggered cAMP-dependent chemotaxis that leads to the formation of multicellular slugs, as the DN mutant of PAKA only delays the onset of this aggregation [26]. However, expression of the kinase domain of PAKA alone, which is constitutively active, blocks slug formation, hinting that both the AID and GBD of PAKA (as well as other three PAKs) could be functionally matured, just like that of multicellular PAKs.

The CRIB sequence of *Dictyostelium* and *Acanthamoeba* PAKs apparently belongs to the unicellular PAKs (the bottom group in Figure 1.6) if judged only by the extent of their overall sequence homology (31–37%) to human PAKs, shown in Figure 1.7.

However, based on the 3D structure of human CDC42 bound to human PAK1-GBD [33], one of us, an expert in the 3D structure of PAKs at Los Alamos National Laboratory, has predicted that the CRIB motif in *Dictyostelium* PAKs may have a significantly weaker affinity for GTPases.

Of the 16 amino acids that comprise the CRIB motif, 6 residues (I75, L77, P78, F81, H83, and H86) are mainly involved in the interaction with GTPase. The contribution of these residues to the GTPase–CRIB interaction can be ranked as follows:

Rank 1: F81
Rank 2: H83
Rank 3: H86
Rank 4: P78
Rank 5: L77
Rank 6: I75

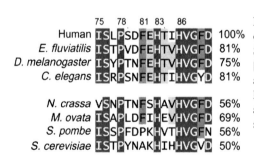

Figure 1.6 CRIB sequence alignment of PAKs from vertebrates to yeasts. Simplified version of figure 6A in Ref. [32]. The CRIB sequence is identical among vertebrate (from human to fish) PAKs. Overall sequence homology (%) against human CRIB (of PAK1–3) is shown on the right.

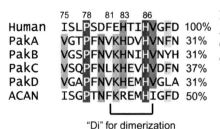

Figure 1.7 CRIB sequence alignment of *Dictyostelium* and *Acanthamoeba* PAKs. Overall sequence homology (%) against human CRIB (of PAK1–3) is shown on the right.

In the alignment in Figure 1.7, the Rank 1 and Rank 5 residues are very different for the *Dictyostelium* PAKs. F81 is replaced by smaller hydrophobic residues in *Dictyostelium* PAKs (A–D), which would weaken their interaction with GTPases. Similarly, L77 is mutated to a polar residue in *Dictyostelium* PAKA–C, or Ala in PAKD. In either case, the change at position L77 results in a loss of affinity. In the absence of major contributions from F81 and L77, GTPases bind more weakly to the *Dictyostelium* PAKs CRIB. Hence, it is more difficult to release the AID from the kinase domain, and it is unlikely that the *Dictyostelium* PAKs would be oncogenic if expressed in mammalian fibroblasts.

In the case of *Acanthamoeba* PAK (Figure 1.7), all of the top-ranked residues are present, except for L77, which is replaced by Gly (expected to reduce its affinity for GTPases) and H83, which is replaced by Arg. Position 83 is in proximity to the negatively charged Glu31 and Asp38 of CDC42 (Figure 1.8A). Since Arg is more positively charged than His, it is predicted that the *Acanthamoeba* CRIB may have a much tighter interaction with GTPases. Furthermore, Arg has a longer side chain and can contribute to an additional link between the CRIB motif and GTPases (Figure 1.8B). As a result of the tighter interaction with GTPases, the interaction of the *Acanthamoeba* AID with the kinase domain would be weakened, in particular because residues 81–87 (the Di motif) are responsible for homodimerization of PAK1 as well (Figure 1.7). In this context, it should also be worth noting that E82 in this Di motif of PAK is invariably conserved in all metazoans, from the freshwater sponges to humans, to form the *trans* double salt bridges with H86 in the antiparallel homodimer, whereas in *Dictyostelium* and *Acanthamoeba* PAKs E82 is replaced by Lys, which clearly blocks the tight double salt bridge formation between positions

(A)   (B)

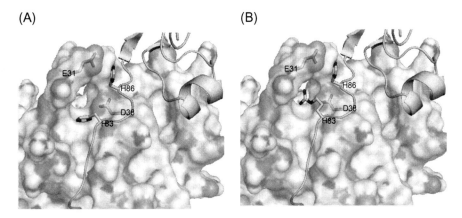

Figure 1.8 GTPase binding to the CRIB motif of PAK1. (A) The wild type of PAK1 (H83) bound to CDC42 [33]. (B) The R83 mutant of PAK1 bound to CDC42, predicted by a molecular modeling. Green surface, CDC42; cyan cartoon, PAK1 CRIB motif (H83 or R83). (For interpretation of the references to color in this figure legend, the reader is referred to the web version of this book.)

82 and 86 (Figure 1.9), suggesting that homodimerization of these amoeba PAKs is rather loose compared with that in metazoan PAKs.

Thus, it is most likely that the *Acanthamoeba* PAK could potentially be oncogenic, similar to the *M. ovata* PAK. This prediction can be tested directly by expression in normal mammalian fibroblasts of *Dictyostelium* or *Acanthamoeba* PAKs, or a mutant of mammalian PAK1 carrying the *Acanthamoeba* Di motif. It is worth noting that unlike the CRIB motif sequence, the sequence of CDC42/RAC is highly conserved from human to yeasts.

1.4 PIX in Fruit Fly *Drosophila* and Nematode *C. elegans*

If we use the degree of sequence homology of GBD/AID in PAKs as a rather rough but convenient scale to measure the evolutionary hierarchy among PAKs, human PAK1 looks closely related to the PAK1 of *C. elegans*, the freshwater sponge, hydra and *Drosophila* (Figure 1.9). All of these metazoan PAKs also contain a PIX-binding motif [32].

```
Hs Pak1  186 PPVIAPRPEHTKSVYTRS 18 (100%)
Dm Pak1  285 PPPVASRPERTKSIYTRP 12 (67%)
Ce Pak1  179 PPAIPDRPARTLSIYTKP 9 (50%)
```

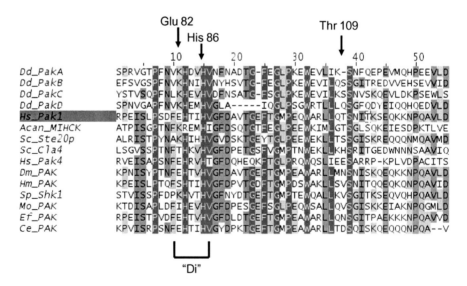

Figure 1.9 GBD/AID sequence alignment of PAKs. Thr 109 of human (Hs) PAK1 corresponds to the residue Thr 38 next to the highly conserved Ser 39 in the above alignment. This potentially LKB1-phosphorylatable Ser/Thr is conserved only in PAKB and PAKC among *Dictyostelium* (Dd) PAKs. The positions equivalent to Glu 82 and His 86 in the Di motif of Hs PAK1 that form a trans salt bridge during dimerization are indicated by arrows.

PIX-dependent activation of PAKs evolved very early in the metazoans. Thus, it is conceivable that PIX is also essential for controlling cell–cell interaction or cell adhesiveness in multicellular animals.

How far can we track PIX on an evolutionary tree of the animal kingdom? A group at the University of California (Davis) found that in *C. elegans* PIX forms a complex with PAK1 (but not with two other PAKs, namely PAK2 and MAX-2) and a scaffolding protein called GIT (G protein-coupled receptor kinase interactor), independently of RAC/CDC42, to control the cell shape and migration of the distal tip cells (DTCs) during morphogenesis of the gonad [34]. In addition, this RAC/CDC42-independent PAK1 pathway functions in parallel to a classical RAC/CDC42-PAK1 pathway to control the guidance aspect of DTC migration. More interestingly, as discussed in Chapter 7, a PAK1-null mutant of this worm is quite healthy (although its litter size is far smaller than the wild type's) and can live significantly longer than the wild type, suggesting that this PIX-PAK1-GIT complex is not absolutely essential for the embryogenesis of this worm, but does have a specialized role in reproductive function.

Unlike the nematode, which keeps blindly crawling on the surface, the fruit fly has big eyes and powerful wings as well as six legs to walk upon. Thus, the fruit fly appears to stand at a far higher level than the nematode on the evolutionary tree of the animal kingdom. As expected, *Drosophila* contains a PIX [35]. DmPIX is required for the stability of DmPAK1 and plays a major role in regulating postsynaptic structure and protein localization at the glutamatergic neuromuscular junction. Loss-of-function mutations of DmPIX lead to a complete inactivation of DmPAK1, confirming that *Drosophila* PIX has a functional role very similar to that of human PIX.

Thus, somewhere between *Dictyostelium* or *M. ovata* and *C. elegans*, both the PIX ancestor and its binding motif on PAK1–3 must have evolved. The nematode appears to be less sophisticated than the freshwater sponge and hydra among the invertebrates whose PAKs bear the PIX-binding motif, based on the following recent observations.

Like the nematode, the tiny freshwater hydra has no eyes. However, a hydra will contract into a ball when exposed to a sudden bright light. David Plachetzki's group [36] at the University of California, Santa Barbara found that hydras "see" light using two proteins similar to those in our own eyes. Rod and cone cells in the human retina contain the membrane-bound G protein-coupled receptor family called opsins, which change their shape when light strikes them. This causes another protein, the cAMP-gated (CNG) ion channel, to generate an electrical signal along nerves connecting the eye to the brain—a process called phototransduction. Hydras have both opsins and ion channels, just as vertebrates do, and make them together in nerve cells. Moreover, a drug called *cis*-diltiazem, which blocks those channels, prevents hydras from responding to light [36]. The sponge larva also can respond to light. Does *C. elegans* respond to light? Well, the blind nematode has no photoreceptors comparable to the opsin system, but it does have a CNG ion channel. If the vertebrate opsin is expressed transgenically in this worm, the worm is able to "see" (respond to) light, changing its locomotive behavior [37]. In other words, in terms of phototransduction, both hydras and sponges appear to be one more step more mature (sophisticated) than nematodes.

1.5 FYN/ETK-Dependent Activation of PAK1

In RAS-transformed mammalian cells, PAK1 requires at least two distinct Tyr kinases, FYN and ETK, for full activation [38–41]. These Tyr kinases appear to work independently. Treatment of RAS cancer cells with PP1/PP2 (FYN inhibitors) alone at 10 nM–10 µM blocks both the kinase activity of PAK1 and the anchorage-independent (malignant) growth of cancer cells by 50% [38,39], whereas treatment with AG 879 (ETK inhibitor) alone at 5 nM–10 µM causes basically the same effect on both PAK1 and malignant growth [17,35]. However, a combination of both inhibitors at 10 nM completely blocks both PAK1 activation and malignant growth [17]. Interestingly, RAS activates both *FYN* and *ETK* genes [39,40], and *FYN*-null mice are basically healthy, but up to 50% less fertile than the control counterparts [41]. This is probably due to the fact that FYN contributes to PAK1 activation only by 50%. As discussed in detail in Chapter 7, a *PAK1*-null nematode is also healthy, but far less fertile (only 14% as fertile) as its control counterparts (and lives significantly longer than they do!). Furthermore, Eric Kandel's group at Columbia University found that FYN is required for long-term potentiation (LTP) of memory function in the hippocampus of adult mice [42].

How can FYN induce LTP? FYN phosphorylates the NR2A and NR2B subunits of the NMDA (*N*-methyl-D-aspartate) receptor [43]. This receptor is a prominent ligand-gated and voltage-gated ion channel for excitatory synaptic transmission in the mammalian central nervous system. Phosphorylation of NR2B at Tyr-1472 by FYN blocks the internalization of this receptor, and thereby keeps it constitutively active. It still remains to be clarified how FYN-induced activation of the NMDA receptor leads to the activation of one of the group I PAKs that have been shown to be involved in LTP. For the details of the role of PAKs in LTP, see the section on Alzheimer's disease in Chapter 6.

What about the situation in the lower eukaryotes, such as nematodes and fruit flies?

C. elegans expresses an *FYN*-related gene called SRC-2 [44]. In 2010, a group at the University of Paris found that *FYN* is required for the human prion protein (PrP) to cause neurotoxicity in the worm. The longevity modulator Sir-2.1/SIRT1, an NAD-dependent histone deacetylase (HDAC), blocks the prion's neurotoxicity [45].

Resveratrol, a polyphenol known to act through sirtuins for neuroprotection, reversed mutant PrP's neurotoxicity in a Sir-2.1–dependent manner. Additionally, resveratrol reversed cell death caused by mutant PrP in cerebellar granule neurons from prnp-null mice. These results suggest that FYN mediates mutant PrP's neurotoxicity.

Since FYN is responsible for 50% of the activation of PAK1, it is conceivable that PAK1 is also involved in prion-dependent neuronal diseases. In this context, it should be noted that resveratrol is known to block PAK1 by upregulating the tumor suppressor PTEN, a PIP3 phosphatase [44]. Furthermore, like FYN-null mice, a PAK1-null strain (RB689) of this worm is healthy but less fertile (80–90% reduction in fertility) and lives significantly longer than the control N2, as discussed in Chapter 7. Thus, it would be of great interest to test whether RB689 is resistant to this mutated prion.

1.6 Evolution of Merlin, Nischarin, and LBK1 (PAK Inhibitors)

1.6.1 Merlin

So far three distinct tumor suppressors—merlin, nischarin, and the kinase LKB1—are known to inhibit PAK1 directly in mammals [46–48]. Merlin is the product of the neurofibromatosis type 2 (NF2) gene, and its overexpression blocks RAS transformation [49]. In NF2 patients, dysfunction of merlin causes two distinct brain tumors, meningioma and schwannoma [50]. Merlin-null (homozygous) mice are embryonic lethal, whereas heterozygous mutants are viable but suffer from NF2, with either meningioma or schwannoma, or often both, when the remaining NF2 gene locus is also lost during an early stage of development. Phosphorylation of merlin at Ser 518 by two distinct kinases, PAK1 and PKA, abolishes its tumor suppressor activity [51,52]. PAK1 is abnormally activated in NF2-deficient tumors, and is required for their growth [46].

Merlin is present in *Drosophila*, where it is required for the regulation of cell proliferation and differentiation in both eyes and wings [53]. Also, it has been shown to be involved in male fertility and is essential for morphogenesis of mitochondria during sperm formation [54]. In *Drosophila*, there are at least four distinct isoforms of the PAK family, PAK1–3, and SLIK (SLK/LOK-like kinase). Like mammalian PAK1, *Drosophila* SLIK phosphorylates Merlin at Thr 616 to inactivate it. Interestingly, a merlin-related protein called moesin is also phosphorylated at Thr 559, but activated by SLIK [55]. Furthermore, a *Drosophila* protein phosphatase 1 (called PP1 or flapwing) de-phosphorylates both merlin and moesin for activation and inactivation, respectively [56]. Thus, the flapwing dysfunction results in inactivation of merlin and activation of moesin (Figure 1.10), eventually leading to a so-called "flapper" phenotype that is unable to fly because of defects in the indirect flight muscles [57].

SLIK was originally identified in a misexpression screen by its oncogenicity: it causes overgrowth, including rough eyes, when expressed ectopically in the imaginal disc epithelium. A SLIK-null mutant has a growth defect that reduces the size of larvae by 60–70% [58]. In short, SLIK and merlin appear to antagonize each other in *Drosophila*. However, it remains to be clarified whether *Drosophila* merlin, like

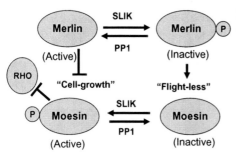

Figure 1.10 *Drosophila* SLIK and PP1 (flapwing) regulating merlin and moesin.

its mammalian counterpart, inhibits the potentially oncogenic SLIK or other PAKs. Unlike conventional PAKs, the SLIK catalytic domain is in the N-terminal half, whereas the C-terminal half bears a few regulatory motifs but no GBD. Nevertheless, SLIK shares some common substrates such as merlin and RAF with mammalian PAKs [51,55,58], induces proliferation, and blocks apoptosis. SLIK acts via RAF but not via the canonical ERK pathway. Activation of RAF can compensate for the lack of SLIK and support cell survival, but activation of ERK cannot [58]. *Drosophila* SLIK is structurally similar to mammalian SLK.

However, unlike that of *Drosophila* SLIK, overexpression of mammalian SLK in cultured fibroblasts results in apoptosis; this shows that the two kinases are functionally dissimilar [59].

1.6.2 Nischarin

Nischarin is a rather pleiotropic protein that binds a few distinct signal transducers such as PAK1, LIM kinase, and integrin [47,60,61]. The N-terminus of nischarin preferentially binds the C-terminal domain of PAK1 only when the kinase is in its activated conformation [49]. Nischarin binding to PAK1 is enhanced by RAC, forming a tertiary complex. Expression of the alpha5beta1 integrin also increases the association of nischarin with PAK1. Nischarin strongly inhibits the kinase activity of PAK1 and cell migration. Furthermore, Nischarin colocalizes with PAK1 in membrane ruffles. Thus, nischarin may regulate cell migration by forming an inhibitory complex with PAK-family kinases. However, whether nischarin is present in eukaryotes other than mammals is not known.

1.6.3 LKB1

LKB1 is an evolutionarily conserved tumor-suppressive kinase present in human *Drosophila* and *C. elegans* [62]. As discussed in Chapter 3, LKB1 is activated when the cellular glucose level drops due to calorie restriction or fasting. LKB1 phosphorylates at least two kinases, AMPK and PAK1 [48]. The tumor-suppressive AMPK is activated by phosphorylation at Thr 172, whereas the oncogenic PAK1 is inactivated by phosphorylation at Thr 109 within the GBD/AID [48]. The importance of LKB1 is highlighted by the fact that it is mutated in a high proportion of Peutz-Jeghers syndrome (PJS) patients. PJS is associated with the development of benign hamartomas as well as a variety of malignant tumors, including nearly half of non-small-cell lung cancers [63,64]. In LKB1-deficient tumors PAK1 is likely to be hyperactivated, and as in merlin-deficient NF2 tumors, may be required for their growth. In mice, deletion of LKB1 causes severe deficits in vascular and neural development that result in embryonic lethality at E10–11 [65,66]. The neural deficit could be in part due to hyperactivation of PAK1–3, which is known to impair the LTP of memory in mice, as discussed in Chapter 5.

  C. elegans LKB1 is called PAR4, and is one of the six "partitioning" molecules that control zygote polarity [67]. The *PAR* genes (PAR1–6) were discovered in genetic screens for regulators of cytoplasmic partitioning in the early embryo of

C. elegans and are required for asymmetric cell division by the worm zygote. Some of the PARs localize asymmetrically and form physical complexes with one another. PAR4 forms a complex with PAR1, a Ser/Thr kinase that is structurally similar to mammalian SAD/MARKs (microtubule-affinity regulating kinases) that phosphorylate microtubule-associated proteins. In the presence of CDC42, alpha-PKC/PAR6 phosphorylates PAR1, which eventually establishes actomyosin-based polarity in the one-cell embryo to determine the anteroposterior axis of the developing worm [67,68]. Strikingly, the PARs, including PAR4/LKB1, regulate cell polarization and asymmetrical cell division in all multicellular animals, suggesting that they form part of an ancient and fundamental mechanism for cell polarization. The lessons learned from *C. elegans* PARs should improve our understanding of how mammalian cells become polarized and divide asymmetrically during development [69].

Drosophila LKB1 is essential for mitotic spindle formation and also affects the asymmetric division of neuroblast stem cells [70]. LKB1 and the other PARs are not absolutely essential for neonatal symmetric cell division in metazoans, although the loss of LKB1 causes malignant tumor formation in mammals. As discussed in Chapters 3 and 7, LKB1 activates FOXO, a tumor-suppressive transcription factor essential for longevity in all metazoans, by either activating AMPK or inactivating PAK1. Thus, loss of LKB1 could shorten the lifespan of this worm as well as mammals, in part by causing cancers, type 2 diabetes, and a variety of other PAK1-dependent diseases.

Finally, it is worth noting that LKB1 is present in *Dictyostelium* [71]. SiRNA-based LKB1 knockdown results in a severe reduction in prespore cell differentiation and a precocious induction of prestalk cells, a phenotype reminiscent of cells lacking GSK3. It remains to be clarified whether or not the anticipated hyperactivation of PAKs is also responsible in part for the deficits in LKB1-null slugs of *Dictyostelium*. Interestingly, among the four *Dictyostelium* PAKs, only PAKB and PAKC share with human PAK1 the potentially LKB1-phosphorylatable site in the AID/GBD (Figure 1.9). Nevertheless, it is clear that LKB1 is essential for cell differentiation or development in all multicellular animals, from humans to this "social" cellular slime mold.

1.7 Concluding Remarks

Although the majority of PAK research has been conducted with mammalian PAKs, mainly because of their relevance to human diseases and their potential as targets for therapeutics, inspiring and valuable studies have been carried out on PAKs in more "primitive" animals such as unicellular amoebas, sponges, nematodes, fruit flies, and even fishes. As seen in this chapter, the latter studies provide us with insights into how mammalian PAKs, in particular their regulatory domains, have functionally evolved to allow us to adapt to surrounding, often harsh environments (such as ice ages and global warming) since *Homo sapiens* emerged in Africa and migrated out of the continent some 50,000–100,000 years ago. The first critical evolution in PAKs occurred during the transition from unicellular to multicellular organisms by mutations in the CRIB motif. In other words, mammalian PAK6, which lacks the

AID, has features typical of some unicellular PAKs. For the phylogeny of the GBD of PAKs, see Figure 1.11. Like MoPAK, PAK4 (and probably PAK6) is potentially oncogenic (for the details, see Chapter 8). Although PAK5 has an AID that is basically identical to that of PAK1, and PAK4 has a newly identified unique AID (residues 20–68) [71], they lack the PIX-binding motif. Thus, PAK4 and PAK5 are similar to early multicellular PAKs that do not require PIX for activation, such as those present in *Dictyostelium*.

PAK1–3 bear both an AID and a PIX-binding motif, and appear to require two Tyr kinases, FYN and ETK, for full activation. Thus, PAK1–3 are among the most functionally mature (or highly sophisticated) members of the PAK family. Surprisingly, however, mammalian PAK3 was found to be expressed in four distinct alternatively spliced variants based on extra exons b or c, which are located between exons 2 and 3 (codons 92 and 93) of the major PAK3 variant called PAK3a [72]. The PAK3b carrying exon b is present in all tetrapods (but not in fish), whereas PAK3c bearing exon c is present only in mammals. The exons b and c encode extra 21 and 15 amino acids, respectively, and these two distinct inserts interrupt both GBD and AID just outside of the CRIB motif (Figure 1.12). More interestingly, unlike PAK3a, these three distinct minor variants (PAK3b, PAK3c, and PAK3bc) are CA forms and show far weaker affinity for GTPases (RAC/CDC42) than does PAK3a [73]. In other words, these CA variants of PAK3 are basically GTPase-independent unicellular- (immature-) type PAKs, and could be potentially oncogenic. Thus, it is conceivable

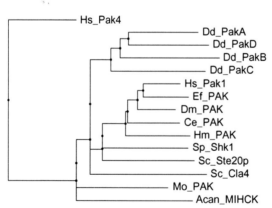

Figure 1.11 Phylogeny of GBD of PAKs. Acan, *Acanthamoeba castellani*; Ce, *C. elegans*; Dd, *Dictyostelium discoideum*; Dm; *Drosophila melanogaster*; Ef, freshwater sponge; Hs, human; Hm, *Hydra magnipapillata*; Mo, *Monosiga ovata*; Sc, *S. cerevisiae*; Sp, *S. pombe*.

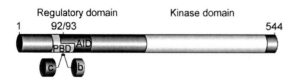

Insert b: **PDLYGSQMCPGKLPE**
Insert c: **NSPFQTSRPVTVASSQSEGKM**

Figure 1.12 Alternatively spiced variants of mammalian PAK3. Schematic structure of PAK3 variants carrying inserts b and c in GBD/AID is shown based on Figure 2A of Ref. [72], which was kindly provided by Dr. Jean-Vianney Barnier.

that during a long evolution, expression of these potentially harmful PAK3 variants could have been subdued stepwise, although they are still expressed in adult mammalian brain. Moreover, PAK3-null mice are both healthy and fertile, although PAK3 deletion appears to cause a significant LD [74].

PAK1 is not essential for the embryogenesis of mice and nematodes, provided that all of the other PAKs remain to compensate functionally for its loss. More interestingly, deregulation (hyperactivation) of PAK1 is now known to cause a variety of human diseases such as cancer and CNS-associated disorders, suggesting that if in the far distant future the *PAK1* gene were to disappear from the human genome, or else be silenced, we could avoid these horrible PAK1-dependent diseases. Moreover, as discussed in Chapter 7, PAK1 is responsible for shortening lifespan in the nematode. The development or identification of effective and safe (and ideally inexpensive) PAK1-blockers in a not-too-distant future could therefore make a major contribution to our fight against these terrible diseases and aging.

Acknowledgment

We are grateful to Mr. Jiri Bohdal for kindly providing us with a photo of the freshwater sponge (Figure 1.5), and Dr. Jean-Vianney Barnier for the structure of PAK3 variants (Figure 1.12).

References

[1] Manser E, Leung T, Salihuddin H, Zhao ZS, Lim L. A brain serine/threonine protein kinase activated by Cdc42 and Rac1. Nature 1994;367:40–6.

[2] Leberer E, Dignard D, Harcus D, Thomas D, Whiteway M. The protein kinase homologue Ste20p is required to link the yeast pheromone response G-protein beta gamma subunits to downstream signalling components. EMBO J 1992;11:4815–24.

[3] Leberer E, Wu C, Leeuw T, Fourest-Lieuvin A, et al. Functional characterization of the Cdc42p binding domain of yeast Ste20p protein kinase. EMBO J 1997;16:83–97.

[4] Kumar A, Molli P, Pakala S, Nguyen T, Rayala S, Kumar R. PAK thread from amoeba to mammals. J Cell Biochem 2009;107:579–85.

[5] Handley M, Rasaiyaah J, Chain BM, Katz D. Mixed lineage kinases (MLKs): a role in dendritic cells, inflammation and immunity?. Int J Exp Pathol 2007;88:111–26.

[6] Harper J, Adami G, Wei N, Keyomarsi K, et al. The p21 Cdk-interacting protein Cip1 is a potent inhibitor of G1 cyclin-dependent kinases. Cell 1993;75:805–16.

[7] El-Deiry W, Tokino T, Velculescu V, Levy D, et al. WAF1, a potential mediator of p53 tumor suppression. Cell 1993;75:817–25.

[8] Parrini M, Lei M, Harrison S, Mayer B. Pak1 kinase homodimers are autoinhibited in *trans* and dissociated upon activation by Cdc42 and Rac1. Mol Cell 2002;1:73–83.

[9] Maruta H, Korn E. *Acanthamoeba* cofactor protein is a heavy chain kinase required for actin activation of the Mg^{2+}-ATPase activity of *Acanthamoeba* myosin I. J Biol Chem 1977;252:8329–32.

[10] Brzeska H, Szczepanowska J, Hoey J, Korn E. The catalytic domain of *Acanthamoeba* MIHCK. II. Expression of active catalytic domain and sequence homology to PAK. J Biol Chem 1996;271:27056–27062.

[11] Lee SF, Egelhof T, Mahasneh A, Cote G. Cloning and characterization of a *Dictyostelium* myosin I heavy chain kinase activated by CDC42 and RAC. J Biol Chem 1996;271:27044–27048.

[12] Brzeska H, Young R, Knaus U, Korn E. Myosin I heavy chain kinase: cloning of the full-length gene and acidic lipid-dependent activation by Rac and Cdc42. Proc Natl Acad Sci USA 1999;96:394–9.

[13] Brzeska H, Young R, Tan C, Szczepanowska J, Korn E. Calmodulin-binding and autoinhibitory domains of *Acanthamoeba* myosin I heavy chain kinase, a p21- activated kinase (PAK). J Biol Chem 2001;276:47468–47473.

[14] Hammer III JA, Seller J, Korn E. Phosphorylation and activation of smooth muscle myosin by *Acanthamoeba* MIHCK. J Biol Chem 1984;259:3224–9.

[15] Manser E, Loo T, Koh C, Zhao Z, et al. PAK kinases are directly coupled to the PIX family of nucleotide exchange factors. Mol Cell 1998;1:183–92.

[16] Obermeier A, Ahmed S, Manser E, Ye SC, et al. PAK promotes morphological changes by acting upstream of Rac. EMBO J 1998;17:4328–39.

[17] He H, Hirokawa Y, Manser E, Lim L, et al. Signal therapy for RAS induced cancers in combination of AG 879 and PP1, specific inhibitors for ErbB2 and Src family kinases, that block PAK activation. Cancer J 2001;7:191–202.

[18] Brzeska H, Kulesza-Lipka D, Korn E. Inhibition of *Acanthamoeba* MIHCK by Ca^{2+}-calmodulin. J Biol Chem 1992;267:23870–23875.

[19] Novak K, Titus M. The myosin I SH3 domain and TEDS rule phosphorylation site are required for in vivo function. Mol Biol Cell 1998;9:75–88.

[20] Brzeska H, Lynch T, Martin B, Korn E. The localization and sequence of the phosphorylation sites of *Acanthamoeba* myosins I. An improved method for locating the phosphorylated amino acid. J Biol Chem 1989;264:19340–19348.

[21] Bement W, Mooseker M. TEDS rule: a molecular rationale for differential regulation of myosins by phosphorylation of the heavy chain head. Cell Motil Cytoskeleton 1995;31:87–92.

[22] Lee SF, Mahasneh A, De La Roche M, Côté G. Regulation of the PAK-related *Dictyostelium* myosin I heavy chain kinase by auto-phosphorylation, acidic phospholipids and Ca^{2+}-calmodulin. J Biol Chem 1998;273:27911–27917.

[23] De La Roche M, Mahasneh A, Lee SF, Rivero F, Côté G. Cellular distribution and functions of wild-type and constitutively activated *Dictyostelium* PakB. Mol Biol Cell 2005;16:238–47.

[24] Falk D, Wessels D, Jenkins L, Pham T, Kuhl S, Titus M, et al. Shared, unique and redundant functions of three members of the class I myosins (MyoA, MyoB and MyoF) in motility and chemotaxis in *Dictyostelium*. J Cell Sci 2003;116:3985–99.

[25] Lee S, Rivero F, Park KC, Huang E, Funamoto S, et al. *Dictyostelium* PAKc is required for proper chemotaxis. Mol Biol Cell 2004;15:5456–69.

[26] Chung CY, Firtel RA. PAKa, a putative PAK family member, is required for cytokinesis and the regulation of the cytoskeleton in *Dictyostelium discoideum* cells during chemotaxis. J Cell Biol 1999;147:559–76.

[27] De La Roche MA, Smith JL, Betapudi V, Egelhoff TT, Côté GP. Signaling pathways regulating *Dictyostelium* myosin II. J Muscle Res Cell Motil 2002;23:703–18.

[28] Azarnia R, Reddy S, Kmiecik T, Shalloway D, Loewenstein W. The cellular src gene product regulates junctional cell-to-cell communication. Science 1988;239:398–401.

[29] De Feijter A, Trosko J, Krizman D, Lebovitz R, Lieberman M. Correlation of increased levels of Ha-ras T24 protein with extent of loss of gap junction function in rat liver epithelial cells. Mol Carcinog 1992;5:205–12.

[30] Nwariaku F, Liu Z, Zhu X, Nahari D, Ingle C, Wu RF, et al. NADPH oxidase mediates vascular endothelial cadherin phosphorylation and endothelial dysfunction. Blood 2004;104:3214–20.

[31] Segawa Y, Suga H, Iwabe N, Oneyama C, Akagi T, Miyata T, et al. Functional development of Src tyrosine kinases during evolution from a unicellular ancestor to multicellular animals. Proc Natl Acad Sci USA 2006;103:12021–12026.

[32] Watari A, Iwabe N, Masuda H, Okada M. Functional transition of Pak proto- oncogene during early evolution of metazoans. Oncogene 2010;29:3815–26.

[33] Morreale A, Venkatesan M, Mott H, Owen D, Nietlispach D, Lowe P, et al. Structure of Cdc42 bound to the GTPase binding domain of PAK. Nat Struct Biol 2000;7:384–8.

[34] Lucanic M, Cheng HJ. A RAC/CDC-42-independent GIT/PIX/PAK signaling pathway mediates cell migration in C. elegans. PLoS Genet 2008;4:e1000269.

[35] Parnas D, Haghighi A, Fetter R, Kim SW, Goodman C. Regulation of postsynaptic structure and protein localization by the Rho-type guanine nucleotide exchange factor dPix. Neuron 2001;32:415–24.

[36] Plachetzki D, Fong CR, Oakley T. Cnidocyte discharge is regulated by light and opsin-mediated phototransduction. BMC Biol 2012;10:17.

[37] Cao P, Sun W, Kramp K, Zheng M, Salom D, Jastrzebska B, et al. Light- sensitive coupling of rhodopsin and melanopsin to G(i/o) and G(q) signal transduction in C. elegans. FASEB J 2012;26:480–91.

[38] He H, Hirokawa Y, Levitzki A, Maruta H. An anti-Ras cancer potential of PP1, an inhibitor specific for Src family kinases: in vitro and in vivo studies. Cancer J 2000;6:243–8.

[39] Yadav V, Denning MF. Fyn is induced by Ras/PI3K/Akt signaling and is required for enhanced invasion/migration. Mol Carcinog 2011;50:346–52.

[40] He H, Hirokawa Y, Gazit A, Yamashita Y, et al. The Tyr-kinase inhibitor AG879, that blocks the ETK-PAK1 interaction, suppresses the RAS-induced PAK1activation and malignant transformation. Cancer Biol Ther 2004;3:96–101.

[41] Luo JP, McGinnis L, Kinsey W. Role of Fyn kinase in the oocyte developmental potential. Reproduct Fertil Develop 2010;22:966–76.

[42] Kojima N, Wang J, Mansuy I, Grant S, Mayford M, Kandel E. Rescuing impairment of long-term potentiation in fyn-deficient mice by introducing Fyn transgene. Proc Natl Acad Sci USA 1997;94:4761–5.

[43] Nakazawa T, Tezuka T, Yamamoto T. Regulation of NMDA receptor function by Fyn-mediated tyrosine phosphorylation. Nihon Shinkei Seishin Yakurigaku Zasshi 2002;22:165–7.

[44] Waite K, Sinden M, Eng C. Phytoestrogen exposure elevates PTEN levels. Hum Mol Genet 2005;14:1457–63.

[45] Bizat N, Peyrin J, Haïk S, Cochois V, Beaudry P, Laplanche J, et al. Neuron dysfunction is induced by prion protein with an insertional mutation via a Fyn kinase and reversed by sirtuin activation in C. elegans. J Neurosci 2010;30:5394–403.

[46] Hirokawa Y, Tikoo A, Huynh J, Utermark T, et al. A clue to the therapy of neurofibromatosis type 2: NF2/merlin is a PAK1 inhibitor. Cancer J 2004;10:20–6.

[47] Alahari S, Reddig P, Juliano R. The integrin-binding protein Nischarin regulates cell migration by inhibiting PAK. EMBO J 2004;23:2777–88.

[48] Deguchi A, Miyashi H, Kojima Y, Okawa K, et al. LKB1 inactivates PAK1 by phosphorylation of Thr 109 in the p21-binding domain. J Biol Chem 2010;285:18283–18290.

[49]   Tikoo A, Varga M, Ramesh V, Gusella J, et al. An anti-RAS function of neurofibromato-
       sis type 2 gene product (NF2/Merlin). J Biol Chem 1994;269:23387–23390.
[50]   Maruta H. Effective NF therapeutics blocking PAK1. Drug Disc Ther 2011;5:266–78.
[51]   Xiao GH, Beeser A, Chernoff J, Testa J. PAK links Rac/Cdc42 signaling to Merlin.
       J Biol Chem 2002;277:883–6.
[52]   Alfthan K, Heiska L, Grönholm M, Renkema G, et al. PKA phosphorylates merlin at
       serine 518 independently of PAK and promotes merlin-ezrin heterodimerization. J Biol
       Chem 2004;279:18559–18566.
[53]   McCartney B, Kulikauskas R, LaJeunesse D, Fehon R. The neurofibromatosis-2 homo-
       logue, Merlin, and the tumor suppressor expanded function together in *Drosophila* to
       regulate cell proliferation and differentiation. Development 2000;127:1315–24.
[54]   Bolobolova E, Iudina O, Dorogova N. *Drosophila* tumor suppressor Merlin is essential
       for morphogenesis of mitochondria during sperm formation. Tsitologiia 2011;53:31–8.
[55]   Hughes S, Fehon R. Phosphorylation and activity of the tumor suppressor Merlin and
       the ERM protein Moesin are coordinately regulated by the slik kinase. J Cell Biol
       2006;175:305–13.
[56]   Yang Y, Primrose D, Leung AC, Fitzsimmons R, McDermand M, Missellbrook A,
       et al. The PP1 phosphatase flapwing regulates the activity of Merlin and Moesin in
       Drosophila. Dev Biol 2012;361:412–26.
[57]   Raghavan S, Williams I, Aslam H, Thomas D, Szöör B, Morgan G, et al. Protein
       phosphatase 1beta is required for the maintenance of muscle attachments. Curr Biol
       2000;10:269–72.
[58]   Hipfner D, Cohen S. The *Drosophila* sterile-20 kinase slik controls cell proliferation and
       apoptosis during imaginal disc development. PLoS Biol 2003;1:E35.
[59]   Sabourin L, Rudnicki M. Induction of apoptosis by SLK, a Ste20-related kinase.
       Oncogene 1999;18:7566–75.
[60]   Alahari SK, Lee JW, Juliano R. Nischarin, a novel protein that interacts with the integrin
       alpha5 subunit and inhibits cell migration. J Cell Biol 2000;151:1141–54.
[61]   Ding Y, Milosavljevic T, Alahari S. Nischarin inhibits LIM kinase to regulate cofilin
       phosphorylation and cell invasion. Mol Cell Biol 2008;28:3742–56.
[62]   Tamás P, Macintyre A, Finlay D, Clarke R, et al. LKB1 is essential for the proliferation
       of T-cell progenitors and mature peripheral T cells. Eur J Immunol 2010;40:242–53.
[63]   Hemminki A, Markie D, Tomlinson I, Avizienyte E, Roth S, Loukola A, et al. A serine/
       threonine kinase gene defective in Peutz-Jeghers syndrome. Nature 1998;391:184–7.
[64]   Sanchez-Cespedes M. A role for LKB1 gene in human cancer beyond the Peutz-Jeghers
       syndrome. Oncogene 2007;26:7825–32.
[65]   Kemphues K, Priess J, Morton D, Cheng NS. Identification of genes required for cyto-
       plasmic localization in early C. elegans embryos. Cell 1988;52:311–20.
[66]   Asada N, Kamon Sanada K, Fukada Y. LKB1 regulates neuronal migration and neuronal
       differentiation in the developing neocortex through centrosomal positioning. J Neurosci
       2007;21:11769–11775.
[67]   Goldstein B, Macara I. The PAR proteins: fundamental players in animal cell polariza-
       tion. Dev Cell 2007;13:609–22.
[68]   Shelly M, Poo MM. Role of LKB1-SAD/MARK pathway in neuronal polarization. Dev
       Neurobiol 2011;71:508–27.
[69]   Lyczak R, Gomes J, Bowerman B. Heads or tails: cell polarity and axis formation in the
       early C. elegans embryo. Dev Cell 2002;3:157–66.

[70] Bonaccorsi S, Mottier V, Giansanti M, Bolkan B, Williams B, Goldberg M, et al. The *Drosophila* Lkb1 kinase is required for spindle formation and asymmetric neuroblast division. Development 2007;134:2183–93.

[71] Bascaran Y, Ng YW, Selamat W, Ling FT, Manser E. Group I and II mammalian PAKs have different modes of activation by CDC42. EMBO Rep 2012;13:653–9.

[72] Veeranki S, Hwang SH, Sun T, Kim B, Kim L. LKB1 regulates development and the stress response in *Dictyostelium*. Dev Biol 2011;360:351–7.

[73] Kreis P, Rousseau V, Thévenot E, Combeau G, Barnier J. The four mammalian splice variants encoded by the p21-activated kinase 3 gene have different biological properties. J Neurochem 2008;106:1184–97.

[74] Meng J, Meng Y, Hanna A, Janus C, Jia Z. Abnormal long-lasting synaptic plasticity and cognition in mice lacking the mental retardation gene Pak3. J Neurosci 2005;25:6641–50.

2 Oncogenicity of PAKs and Their Substrates

Hong He[1], and Hiroshi Maruta[2]

[1]University of Melbourne, Austin Health, Melbourne, Australia, [2]NF/TSC Cure Org, Brunswick West, Melbourne, Australia

Abbreviations

AMPK	AMP-activated kinase
BAD	Bcl-2 antagonist of cell death
CPI17	17-kDa PKC-potentiated inhibitory protein of PP1
DLC1	dynein light chain 1
ER	estrogen receptor
FOXO	forkhead box protein O1
GIT1	G protein-coupled receptor kinase interactor 1
MBS	myosin-binding subunit of type 1 protein phosphatase
MEK1	mitogen-activated protein kinase 1
MEKK1	mitogen-activated protein kinase kinase 1
MLCK	myosin light chain kinase
PIX	PAK-interacting exchange factor
Rho GDI	Rho GDP dissociation inhibitor
R-MLC	regulatory myosin light chain

2.1 Introduction

In general, cancers are caused by a combination of gain-of-function (activating) mutations of protooncogenes such as *RAS* and *SRC* and loss-of-function (dysfunctioning) mutations of tumor-suppressor genes such as *p53* and *APC*[1]. Among these mutations, *p53* mutations and *RAS* mutations are found in more than 50% and 30% of all human cancers, respectively [1,2]. In mammals three distinct isoforms of RAS (Ki, Ha, and N) are expressed, and Ki-RAS mutations occur most frequently in human cancers, notably in 90% of pancreatic cancers and 50% of colon cancers [1]. These RAS isoforms are GTPases (G proteins) whose function depends on GTP and farnysylation or geranylgeranylation at their C-terminus [3]. Thus, a series of compounds (FTIs or GTIs) that block farnysyl-transferase (FT) or geranylgeranyl-transferase (GT) essential for RAS C-terminal lipid modification were developed by several

PAKs, RAC/CDC42 (p21)-activated Kinases. DOI: http://dx.doi.org/10.1016/B978-0-12-407198-8.00002-3

pharmaceutical companies such as Merck during the 1990s. However, these RAS inhibitors turned out to be too toxic for clinical application. Since then, oncogenic effectors downstream of RAS have been explored as the potential targets of new anti-cancer drugs. One of these RAS effectors, called PI-3 kinase, is a unique lipid kinase that phosphorylates PIP2, producing the oncogenic PIP3 [4]. However, since PIP3 is essential for normal cell survival or growth, inhibitors of PI-3 kinase are also too toxic for clinical application. Meanwhile two GTPases, RAC and CDC42, were found to be essential for RAS-induced malignant transformation [5,6]. However, FTIs or GTIs that block the C-terminal lipid modification of RAC or CDC42 would be too toxic for clinical application.

In the mid-1990s, two distinct kinases, called PAK and ACK, respectively, were discovered by Ed Manser's group in Singapore [7,8]. PAKs are RAC/CDC42-activated Ser/Thr kinases [7], whereas ACK is a CDC42-activated Tyr kinase [8]. During 1997 and 1998, Jeff Field's group at the University of Pennsylvania found that PAK1 is essential for the growth of RAS transformants and malignant peripheral nerve sheath tumor (MPNST), a *NF1*-deficient malignant tumor [9,10]. The tumor-suppressor *NF1* gene product is an RAS GAP, which attenuates normal RAS [10], and dysfunction of this GAP causes hyperactivation of RAS. In 1999, we found that ACK is also essential for the growth of RAS transformants [11]. Since then increasing evidence indicates that PAKs, in particular PAK1 and PAK4, are hyper-activated or overexpressed in a variety of human tumors such as pancreatic, colon, breast, prostate, and lung cancers as well as those associated with neurofibromatosis (NF). Once PAK1 or PAK4 is hyperactivated or overexpressed, it activates a number of potentially oncogenic kinases such as RAF, LIM kinase and aurora-A or inacti-vates tumor suppressors such as BAD, merlin, and FOXO, leading to the oncogenic (malignant) transformation of normal cells.

The definition of "malignancy" depends on the conditions where cells grow. *In vitro* (in cell culture), malignancy usually means the ability of cells to grow (pro-liferate) in anchorage- or serum-independent manners. However, *in vivo* (in whole animals or patients), malignancy often means the ability of tumor cells to grow indefinitely or cause metastasis (translocate from one organ to another by circulation or tissue invasion).

Here we shall explain briefly the molecular basis of several criteria for malig-nancy in which PAKs, mainly PAK1 and PAK4, are involved.

2.2   Loss of Contact Inhibition of Growth

Normal fibroblasts or epithelial cells cease to divide/proliferate and move as soon as they contact each other (reach the confluence), forming a so-called "gap junction." Back in the 1950s, Mike Abercrombie (1912–1979) at the University of Cambridge called this unique phenotype (social behavior) of normal cells as "contact inhibi-tion" of growth and movement [12]. However, he found that once cells are malig-nantly transformed by RAS or other oncogenes, they no longer stop their growth and movement even if they reach the confluence [12]. Thus, he chose this loss of contact inhibition as the first criterion for malignancy *in vitro*. If you culture cells on cover slips, normal cells form a monolayer and stop their growth, but malignant cells keep growing, overlapping each other to form cell masses called foci. Thus, foci formation can be used as a criterion for malignancy *in vitro*. RAS blocks contact inhibition, inducing foci formation by activating PAK1, which disrupts gap junction or tight cell–cell adhesion [13]. How does PAK1 disrupt gap junction? The contact inhibition requires the interaction of a cell surface receptor called CD44 with a tumor suppressor called merlin. However, PAK1 phosphorylates merlin [14], and the phospho-merlin forms a heterodimer with an oncogenic protein called ezrin [15]. This complex no longer interacts with CD44. Thus, once PAK1 is hyperactivated, cells no longer form gap junctions and lose contact inhibition.

2.3   Serum-Independent Growth

Most normal cells, except, perhaps, blood cells, require serum for their growth. However, malignant cells can grow without serum. Why and how? Serum contains a specific growth factor called PDGF that binds its cell surface receptor (PDGFR), a Tyr kinase, which transactivates EGF receptors ErbB1 and ErbB2, cell surface Tyr kinases, which in turn transiently activate PAK1 through RAS and other cytoplasmic signal transducers [16]. In other words, normal cells require serum to transiently activate PAK1 for their growth. However, RAS-transformed cells no longer need serum because PAK1 is hyperactivated, and constitutively stimulate the secretion of a few growth factors such as TGF-α and heregulin, which activates the ErbB1/ErbB2 heterodimer [16]. Thus, malignant cells can grow even in serum-free medium.

2.4   Anchorage-Independent Growth

Normal fibroblasts or epithelial cells cannot grow in suspension culture and have to adhere on a substratum for their growth. This phenotype is called "anchorage-dependent" growth. However, malignant cells can grow even in suspension cultures such as soft agar plate. This malignant phenotype is called "anchorage-independent" growth [17]. Why do normal cells fail to grow in suspension? In suspension, the tumor-suppressor PTEN is constitutively activated by the RhoA–ROCK signaling pathway [18], and consequently two oncogenic kinases, PAK1 and AKT, are inactivated. Furthermore, PAK1 is activated through the integrin–FAK–ETK pathway only when normal cells adhere to the substratum/extracellular matrix [19]. Without a basic level of PAK1 and AKT, normal cells cannot grow in suspension. On substratum, however, PTEN is silenced, and basic levels of both PAK1 and AKT are sufficient for the growth of normal cells. However, malignant/cancer cells such as RAS transformants contain constitutively activated PAK1 (and AKT), and RAS silences PTEN. Thus, cancer cells can grow even on soft agar plates in an anchorage-independent manner.

2.5   Tumor-Induced Angiogenesis

Solid tumors require angiogenesis (new blood vessel formation) for their rapid and infinite growth *in vivo*. For without sufficient blood supply (oxygen and nutrients), solid tumors suffer from hypoxia and stop growing or die away via apoptosis. Interestingly, hypoxia upregulates PAK1 [20], which in turn activates the VEGF gene through β-catenin, essential for tumor-induced angiogenesis [21]. Thus, blocking PAK1 could be among the most effective ways for suppressing the growth of all solid tumors.

2.6   Metastasis/Invasion

Metastasis is the most common criterion for malignancy used by clinical oncologists, mainly because metastasis is the ultimate cause of most cancer patients' premature death. Without metastasis, solid tumors can be often removed by either surgery or X-ray irradiation, but once tumors are metastasized, in particular into the lungs, there is no effective way to physically remove these tumors. Systemic administration of conventional chemotherapy agents (DNA/RNA/MT = microtubule poisons) does not block metastasis and is basically useless against the spread of these cancers in the patient's body. However, PAK1 is essential for metastasis because it activates an effector, called LIM kinase, which is required for cell movement/invasion, and leads to the metastasis of solid tumors by controlling actomyosin dynamics through an F-actin severing protein called cofilin [22–24].

For an outline of the above oncogenic roles of PAK1 and PAK4 and their major oncogenic targets, see Figure 2.1.

2.7   PAK1/PAK4-Dependent Solid Tumors

Unlike *RAS* and *SRC* family genes as well as several cell surface Tyr kinases such as ErbB1 (EGF receptor) and ErbB2, so far there is no report for activating mutations

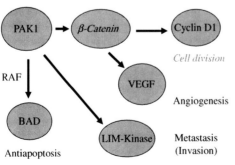

Figure 2.1 Oncogenic roles of PAK1.

of *PAK*1 gene in human cancers. However, several years ago a mutation was found in the kinase domain of *PAK*4 (E329K) in a colorectal tumor sample [25], and in 2012, Claire Wells's group at King's College London demonstrated that this mutation indeed enhances its kinase activity, promoting cell migration [26]. This mutation is the first example among the PAK mutations found in human cancers with the increased kinase activity. Furthermore, hyperactivation, amplification, or overexpression of PAK1 or PAK4 has been found very often in a number of human solid tumors. PAK1 is the isoform most commonly overexpressed, but PAK4 is also overexpressed in specific cancers. PAK4 is overexpressed in 75% of the NCI 60 cell line panel and its dominant negative (DN) mutant, which sequesters either oncogenic substrates or activators of PAK4, blocks effectively the growth of a colon cancer cell line [27].

Both *PAK*1 and *PAK*4 genes are localized to genomic regions called "amplicons," which are frequently amplified in cancer cells. The *PAK*1 gene is localized within the 11q13 region, and 11q13.5-q14 amplifications involving the *PAK*1 locus found in bladder, ovary, and breast cancer [28–30]. *PAK*4 gene localizes to another amplicon, 19q13.2, and its amplification has been found in colorectal and pancreatic cancers [25,31].

*PAK/PAK*4 gene amplifications are not frequent enough to be the only molecular mechanism leading to PAK overexpression in cancer. A recent report identified a novel mechanism for the overexpression of PAK1 through microRNA downregulation. Rakesh Kumar's group [32] found that the levels of endogenous microRNA miR-7 inversely correlated with PAK1 expression in a variety of cancer cell lines. Moreover, miR-7 downregulates PAK1 in breast cancer cells, and suppressed their motility and invasiveness [32].

PAK1 is far more hyperactivated than overexpressed in human cancer cells by a variety of its upstream (oncogenic) activators such as RAS, PI-3 kinase, ETK, FYN, ErbB1 (EGF receptor), and ErbB2 or dysfunction of tumor-suppressor genes such as *NF*1, *NF*2 and *PTEN*, as discussed in detail in Chapter 3.

2.7.1 Pancreatic Cancer

Pancreatic and colon cancers belong to the so-called RAS cancers because more than 90% of the former and 50% of the latter carry oncogenic RAS mutations [2]. Since RAS eventually activates PAK1 through PI-3 kinases and RAC/CDC42 (see Figure 2.2), in these RAS cancers PAK1 is hyperactivated. Just like the growth of colon cancers, discussed later in this chapter, the growth of human pancreatic cancer xenografts in mice was shown to be strongly suppressed by a series of synthetic chemical drugs or natural products such as a combination of two Tyr kinase inhibitors (PP1 and AG 879/Gl-2003), FK228, chloroquine and Bio 30 which is a CAPE (caffeic acid phenethyl ester)-base propolis extract, blocking PAK1 directly or indirectly [33–36]. Furthermore, even in human trials, both terminal (metastasized) and early pancreatic cancers completely disappeared with Bio 30 treatment in a year without any adverse effect [36], clearly indicating that the majority of pancreatic cancers are PAK1 dependent. This is a very encouraging development because the majority of pancreatic cancer patients are expected to survive for only 3–4 months without

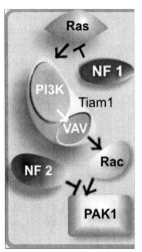

Figure 2.2 RAS to PAK1 signaling cascade.

effective therapeutics, and so far no effective FDA-approved drug is available on the market for these deadly cancers as yet, except for gemcitabine, a DNA poison, to which only 10% or fewer of pancreatic cancer patients respond [37]. For the details of PAK1 blockers, see Chapters 3 and 5.

2.7.2 Colon/Gastric Cancer

As mentioned above, around 50% of human colon cancers carry oncogenic mutations of *RAS* genes, in particular the *Ki-RAS* gene [2]. However, *RAS* mutation alone is not sufficient to cause malignant transformation of colon cells. It also requires a loss-of-function (dysfunction) mutation of a few additional genes such as *APC* and *p53* [38]. Both *APC* and *p53* are tumor-suppressor genes that block RAS-induced malignant transformation [39,40], and dysfunction of adenomatous polyposis coli (APC) and p53 are found in 90% and 70% of human colon cancers, respectively [38].

How does APC block oncogenic RAS signaling? APC forms a complex with an oncogenic transcription factor called β-catenin in cytoplasm that renders β-catenin sensitive to phosphorylation by GSK-3 β or other Ser kinase(s). This leads to a rapid degradation by ubiquitin-based proteosomes [41]. However, in the absence of APC, PAKs (in particular PAK1/PAK4) activated by RAS can phosphorylate a free β-catenin at Ser 675 in the cytoplasm and translocate the β-catenin into the nucleus [42–44], where β-catenin can act as an oncogenic transcription factor that activates the promoters of several genes including *cyclin D*1 and *VEGF*, which are essential for DNA replication and tumor-induced angiogenesis, respectively [45,46].

In addition to RAS-induced hyperactivation of PAK1 in 50% of colon cancers, overexpression of PAK1 is observed in 70% of colon cancers [42], and both are correlated with several signaling pathways including Wnt, ERK, and AKT. The RAC/PAK1 cascade directly controls β-catenin phosphorylation at Ser 675 and its full activation in colon cancer cells [42]. ERK and AKT, downstream targets of PAK1, are also involved in colon cancer progression [43]. PAK1 inhibition alone is equivalent to the dual inhibition of ERK and AKT, whereas inactivation of either the ERK or AKT pathway alone only partially inhibits cell proliferation, migration/invasion, and survival [43]. Thus, instead of simultaneously inhibiting both ERK and AKT, using PAK1 blockers would be a more effective therapeutic for colon cancers. In fact, we demonstrated that PAK1-specific siRNA strongly suppresses the growth of human colon cancer xenograft in mice [44].

As discussed in detail in Chapter 4, the majority of gastric cancers in the East Asian population are caused by infection by a virulent strain(s) of the gastric bacteria *H. pylori* carrying unique CagA variants [47]. Interestingly, this bacterial infection requires PAK1 in host cells [48]. In short, the bacterial protein CagA activates PAK1 in host cells through PIX, a PAK1-activating SH3 adaptor protein, and PIX-specific siRNA blocks the CagA-induced PAK1 activation [48].

Although such a siRNA cannot be used effectively for clinical application, recently a few potent PAK1-specific inhibitors and a PAK1/PAK4 inhibitor (PF-3758309) were developed by Afraxis and Pfizer Oncology, respectively, in San Diego [49], and administration of PF-3758309 (around 20 mg/kg daily) was shown to suppress the growth of several human cancers grafted in mice, including colon cancer [49]. For details of PF-3758309, see Chapter 3.

2.7.3   Breast Cancer

Breast cancer represents around 30% of female cancers, and all breast cancers appear to depend on PAK1 for their growth regardless of their estrogen dependency. More than 50% of human breast cancers display overexpression and/or hyperactivation of PAK1, and PAK1 is found on a chromosomal region amplified in 17% of breast cancers [30,50]. Rakesh Kumar's group [51] at the M.D. Anderson Cancer Center (MDACC) demonstrated that transgenic expression of constitutively active (CA) PAK1 alone in mouse mammary tissue is sufficient to cause breast cancers. More recently Audrey Minden's group [52] at State University of New Jersey showed that PAK4 also promotes tumorigenesis in breast cancer cells. Taken together, these studies clearly indicate the critical role of these PAKs in the development of breast cancers. Several signal transducers, such as RAF/MEK/ERK, Cyclin D1, MET, NFκB, and estrogen receptor α (ERα), are activated by PAK1 during the progression of breast cancer, and these pathways will be discussed in detail below [50–57].

First of all, PAK1 is involved in ER signals in breast cancer. Approximately 75% of all breast cancers express the estrogen receptor (ERα), and tamoxifen, an antagonist of estrogen, is widely used to treat estrogen-dependent breast cancers. PAK1 is one of the major kinases that phosphorylate ERα [54,58]. Hyperactivation of PAK1 produces multiple phosphorylation of ERα, creating a promiscuous receptor that is resistant to tamoxifen (Tam), and stimulates their growth even in the absence of estrogen, namely an estrogen-independent, CA receptor [54,58]. The nuclear levels of active PAK1 increased in breast cancer patients with Tam resistance [54,58]. Moreover, ER activation by PAK1 induces upregulation of cyclin D1 in breast cancer cells, as well as in the mammary epithelium [50]. Independently, we found that RAS-induced upregulation of cyclin D1 also requires PAK1 [59]. Patients with low PAK1 levels obtained more benefit from Tam treatment [58]. The link between PAK1 and ERα raises the possibility that Tam resistance could be reversed by PAK1 inhibition. In fact, we found that PAK1 blockers such as FK228 suppress almost completely the growth of both Tam-resistant and -sensitive breast cancers *in vivo* [60].

PAK1 is activated through pathways that are important for breast cancer growth. Growth factors such as prolactin and the cell surface receptor Tyr kinase ErbB2 can activate PAK1. The prolactin receptor (PRL-R) can initiate and sustain ERK1/2 signaling via the PI-3 kinase-dependent RAC/PAK1 pathway rather than the canonical ErbB2/Shc/Grb2/SOS/RAS route [61]. PRL-R also activates PAK1, leading to the upregulation of cyclin D1 [62]. *ErbB2* gene overexpression, amplification, or mutation occurs in about 25% of human breast cancers [63]. ErbB2 signaling activates a RAC–PAK1 signaling pathway that contributes to ErbB2-mediated transformation through the MAPK/ERK and AKT pathways [64,65]. ErbB2 activates RAC and PAKs in a 3D breast epithelial cell culture system, and the inhibition of RAC or PAK1 activity blocks the morphologic effects of ErbB2 in these cells, accompanied by loss of ERK and AKT activation [64]. Moreover, the kinase activity of PAK1 is required for ErbB2 transformation in a xenograft model of breast cancer [64].

2.7.4 Cervical Cancer and Papillomas

Human papilloma virus (HPV) is among the first identified human oncoviruses (viruses that cause tumors in human beings). By the mid-1980s, Harald zur Hausen's group [66] at the University of Freiburg and other groups had established that HPV is a necessary cause of cervical cancer and a few other genital cancers, and currently a variety of HPV vaccines are available for the therapy of these genital cancers [67]. For this reason, Prof. zur Hausen at DKFZ (the German Cancer Research Center) in Heidelberg (Figure 2.3) received the 2008 Nobel Prize in Physiology or Medicine: http://www.nobelprize.org/nobel_prizes/medicine/laureates/2008/hausen.html.

However, there are 20–60 immunologically distinct types of HPV that regularly or sporadically infect the human genital tract (with type 18 appearing in 50% of cervical cancer biopsies including the HeLa cell line), and therefore the production of vaccines specific for each type would be a very tedious task. The common oncoproteins encoded by all the types of HPV genomes are E6 and E7, which convert normal "mortal" cells to "immortal" ones by blocking their apoptosis (programmed death of senescent cells) [66]. Interestingly, like host cells infected by HIV and flu viruses (as discussed in detail in Chapter 4), papillomas caused by HPV infection were recently shown to carry hyperactivated PAK1 and PAK2, through the ErbB1–RAC signal pathway [68]. In the case of HIV and flu viruses, their infection and replication require PAK1 in host cells. Thus, it is conceivable that HPV infection also requires PAK1/PAK2 in these genital cells. If that were the case, PAK blockers would be potentially far simpler therapeutics for these HPV-dependent cancers as well. In support of this notion, the estrogen-dependent growth of cervical cancer cells was indeed shown to be strongly inhibited by the natural PAK1 inhibitor

Figure 2.3 Harald zur Hausen, 2008 Nobel laureate.

"Curcumin" [69]. For the malignant growth of cervical cancer, these viral oncoproteins (E6 and E7) and the female steroid hormone "estrogen" are known to work synergistically [69], and as discussed in the "ER" section (Section 2.8.5), ER and PAK1 form the vicious oncogenic cycle.

2.7.5 Prostate Cancer

Prostate cancer represents around 30% of male cancers, and all human prostate cancer cell lines (4/4) including PC-3 were first reported to lack the tumor-suppressor PTEN, PIP3 phosphatase [70]. Since PTEN antagonizes PI-3 kinase and suppresses two oncogenic kinases, PAK1 and AKT, downstream of PI-3 kinase, it is anticipated that both PAK1 and AKT are hyperactivated in these PTEN-deficient prostate cancers. Thus, it is conceivable that prostate cancers are among PAK1-dependent cancers. Indeed, the growth of PC-3 is strongly suppressed by FK228, the most potent PAK1 blocker [71]. However, a subsequent more systematic analysis with a larger sample number [70,72] revealed that the incidence of PTEN deficiency (loss of heterozygosity) in highly metastatic prostate cancers is around 70% (14 in 22). More recently, curcumin, which inhibits PAK1 directly, has been found to suppress the growth of human prostate cancer grafted in mice by inhibiting nuclear translocation of beta-catenin, an oncogenic substrate of PAK1 [73].

2.7.6 Neurofibromatosis

Neurofibromatosis types 1 and 2 (NF1 and NF2) are rare genetic diseases or disorders caused by loss-of-function mutations of the tumor-suppressor genes *NF1* and *NF2*, respectively [36]. NF1 is a more common disease, developing in about 1 in 3000 births, while NF2 is a more rare disorder with an incidence of about 1 in 30,000 births. However, NF2 is more severe than NF1 in general because NF2 tumors develop mainly along the spine and in the brain, leading to dysfunction of the central nervous system, whereas 90% of NF1 tumors are benign and developed on the skin. NF patients are predisposed to the development of multiple tumors of the central and peripheral nervous systems. Schwann cells, which comprise the myelin sheath around nerves, are predominantly affected in both tumors. Mice lacking either the *NF*1 or *NF*2 gene (homozygous) are embryonic lethal. Patients can carry heterozygous mutations in either the *NF*1 or *NF*2 gene, but their tumors typically display loss of the residual wild-type allele, conforming to the classic two-hit Knudsen paradigm seen with most tumor suppressors. Although NF1 and NF2 are genetically and clinically distinct diseases, loss of each gene product leads to hyperactivation of PAK1, and PAK1 is essential for tumor growth in both NF1 and NF2 [36].

The *NF*1 gene product is a RAS GAP of 2818 amino acids that attenuates normal RAS by stimulating its intrinsic GTPase activity. Consequently, loss of this GAP leads to the hyperactivation of normal RAS, which activates oncogenic effectors such as PI-3 kinase. PI-3 kinase in turn activates PAKs via RAC and CDC42 (see Figure 2.2). The DN mutants of PAK1 were shown to suppress the growth of human NF1-deficient MPNSTs *in vivo* [10].

We found that the *NF*2 gene product called merlin directly inhibits the kinase activity of PAK1 [74]. Thus, in NF2-deficient tumors such as meningioma, schwannoma, and mesothelioma, PAK1 is hyperactivated, and PAK1 inhibitors such as CEP-1347 and WR-PAK18 were able to block the growth of merlin-deficient tumor cells but not merlin-positive cells [74]. These findings clearly indicate that both NF1 and NF2 tumors require PAK1 for their growth. Thus, in principle, both NF1 and NF2 patients can be effectively treated with PAK1 blockers, as discussed in detail in Chapters 3 and 5.

Among NF tumors, a rare tumor is associated with schwannomatosis (occurring in around 1 in 30,000 births). Although its apparent symptoms are very similar to those of NF2 tumors, the precise genetic cause of schwannomatosis still remains largely unknown, and so far PAK1 blockers appear not to be effective for the therapy of schwannomatosis.

2.7.7 Lung Cancer

Compared with pancreatic, colon, and breast cancers or NF tumors, lung cancers are far more heterologous in terms of oncogenic mutations. Nevertheless, around 30% of human lung cancers, in particular adenocarcinomas, carry an oncogenic *Ki-RAS* mutation [75], emerging as a potential PAK1-dependent cancer. A mouse model for *Ki-RAS*-induced lung cancers requires RAC for their growth, indicating that their growth would depend on its effector PAK1–3 [76]. PAK1 is overexpressed in both nucleus and cytoplasm of squamous nonsmall cell lung carcinomas [30]. Finally, selective inhibition of PAK1, but not PAK2, delays the cell-cycle progression *in vitro* and in *vivo*[30]. As discussed later, more than 50% of human malignant mesothelioma (HMM) developed in the lungs require PAK1 for their growth, mainly because they lack the PAK1 inhibitor merlin. Furthermore, over 10% of human lung cancers carry the oncogenic mutation of the *ErbB*1 (EGF receptor) gene [77]. Since *ErbB*1 activates PAK1 through the RAS–PI-3 kinase–RAC/CDC42 pathway, it is most likely that these *ErbB*1-induced lung cancers also require PAK1 for their growth and can be effectively treated by either *ErbB*1 inhibitors such as Iressa or PAK1 blockers such as FK228 and propolis.

2.7.8 Human Malignant Mesothelioma (HMM)

Human malignant mesothelioma (HMM) is often caused by the exposure of lungs to asbestos. Interestingly, around half of HMMs lack merlin, the NF2 gene product that is a direct inhibitor of PAK1 [74,78]. Thus, in these NF2-deficient HMMs, PAK1 is hyperactivated and their growth is PAK1 dependent [74]. We have shown that the growth of these NF2-deficient HMMs is indeed suppressed by PAK1 blockers such as FK228, CEP-1347, and WR-PAK18 [74]. Unfortunately, however, the remaining 50% of HMMs still express merlin and are resistant to these PAK1 blockers [74].

2.7.9 Ovarian Cancer

A decade ago, by analyzing the amplicon at the chromosomal locus 11q13.5-q14 of human ovarian cancers, a Swiss group in Basel found that the *PAK1* gene is among the most frequently amplified in ovarian cancer cell lines [79]. In this cancer, the DNA copy number gains were most frequently observed for *PIK3CA* on 3q (66%), *PAK1* on 11q (59%), and *KRAS2* on 12p (55%), suggesting the possibility that like pancreatic and colon cancers, ovarian cancers belong to the RAS–PI-3 kinase–PAK1-dependent cancers. Furthermore, the 11q13-q14 amplicon, represented by six onco-genes (*CCND1*, *FGF4*, *FGF3*, *EMS1*, *GARP*, and *PAK1*) revealed preferential gene copy number gains of *PAK1*. *PAK1* copy number gains were observed in 30% of the ovarian carcinomas and overexpression of PAK1 was in 85% of the tumors. *PAK1* gains were associated with high grades ($P < 0.05$). In contrast, *CCND1* gene altera-tions and protein overexpression were less frequent (11% and 25%, respectively). Does ovarian cancer require PAK1 for growth? Yes. During 2009–2010, in collabora-tion with Tamotsu Sudo's group at Hyogo Cancer Center in Japan, we demonstrated that both overexpression and hyperactivation of PAK1 are observed in many human ovarian cancer cell lines, and PAK1 blockers such as TAT-PAK18 and Ivermectin indeed suppress the growth of human ovarian cancer cells, at least *in vitro* [80,81].

2.7.10 Thyroid Cancer

In 2007, Mathew Ringel's group at Ohio State University (OSU) found that a Celecoxib derivative called OSU-03012 inhibits PAK1 directly with the IC_{50} around 5 μM and suppresses the growth of human thyroid cancer cells *in vitro*[82], suggest-ing that thyroid cancers also belong to the PAK1-dependent cancer family. It has been known previously that this compound inhibits another kinase called PDK [83]. However, the IC_{50} for PAK1 is far lower than that for PDK, a PIP3-dependent activa-tor of the kinase AKT, indicating that PAK1 is the primary (and so far the most sensi-tive) target of this compound. It also suppresses the PAK1-dependent growth of human NF2 tumor (Schwannoma) xenografts in mice at a daily oral dose of 200 mg/kg [84].

Interestingly, it was revealed recently that dysfunction of the *NF2* gene product merlin, a PAK1 inhibitor, is a part of the reason why thyroid cancers grow in a PAK1-dependent manner [85].

2.7.11 Multiple Myeloma

In addition to the inhibition of PAK1, OSU-03012 activates AMP-activated kinase (AMPK) in multiple myeloma (MM) cell lines and inhibits MM cell growth in the presence of Gleevec [86], suggesting that myelomas require PAK1 as well as c-Kit for their growth. Furthermore, FK228, a potent HDAC inhibitor that eventually inactivates PAK1, inhibits the growth of MM cells [87]. Moreover, curcumin, which directly inhibits PAK1, inhibits the growth of MM cells [88]. Taken together, these observations clearly indicate that MM is among the PAK1-dependent cancers, although so far neither overexpression nor hyperactivation of PAK1 has been reported.

2.7.12 Gliomas

The growth of several distinct glioma cell lines of human origin is suppressed by a variety of PAK1 blockers such as FK228, CAPE, and curcumin either *in vitro* (cell culture) or *in vivo* (grafted in mice) [89–91]. Furthermore, we have shown that Bio 30, a CAPE-based propolis extract from New Zealand available on the market (100 mg/kg, i.p., twice a week), suppresses strongly the growth of human glioma grafted in mice [36]. These findings indicate that the growth of gliomas in general requires PAK1. However, it is not entirely clear why gliomas' growth is PAK1 dependent. Around 30% (13/42) of brain tumor (glioma) cell lines lack PTEN, a phospholipid (PIP3) phosphatase [70]. Since dysfunction of PTEN causes hyperactivation of two distinct oncogenic kinases, PAK1 and AKT, it is conceivable that these PTEN-deficient gliomas require PAK1 for their growth. For details of gliomas, see Chapter 5.

2.7.13 Hepatocellular Carcinomas

The *PAK1* gene is overexpressed in human hepatocellular carcinomas (HCCs), and its overexpression is associated with more aggressive tumor behavior in terms of more metastatic phenotypes and more advanced tumor stages [92]. In addition, a HCC cell line stably expressing PAK1 displayed increased cell motility rates and, conversely, knockout of endogenous *PAK1* expression by siRNA reduced the migration rates of HCC cells [92], clearly indicating that metastasis of HCCs requires PAK1. Furthermore, curcumin, a PAK1 inhibitor, strongly suppresses HCC-induced angiogenesis and HCC growth in mice [93]. Thus, like many other cancers, HCCs require PAK1-dependent angiogenesis for their rapid growth.

2.7.14 Melanoma

Prognosis in patients with uveal melanoma is poor, as approximately half of all tumors metastasize and currently there are no effective treatments for the disseminated disease. A group at Queensland Institute of Medical Research in Australia found that among six mammalian PAKs, only PAK1 is overexpressed in the more invasive cultures of melanoma cell lines, and PAK1-specific siRNA blocks its invasion through matrigel by 80% [94], clearly indicating that elevated levels of PAK1 are associated with invasive potential in uveal melanoma. Furthermore, a Russian group in Moscow demonstrated in 2004 that Ivermectin (3 mg/kg daily), which blocks PAK1, strongly suppresses the growth of a variety of human cancers grafted in mice, including melanoma [95]. These observations indicate that both growth and metastasis of melanoma require PAK1. Why does melanoma grow in a PAK1-dependent manner? A part of the reason is that like NF2 tumors and mesothelioma, a portion of melanoma lacks the tumor-suppressor merlin, the NF2 gene product that directly inhibits PAK1 [74,85].

2.8   Oncogenic Targets of PAKs

2.8.1   PAK Substrates and Their Phosphorylation Sites

According to a few recent reviews [96,97], over 40 distinct proteins have been identified as substrates for PAKs. These substrates are roughly divided into three categories: (1) potentially oncogenic (protoonco) proteins such as RAF, LIM kinase, and β-catenin; (2) tumor-suppressor proteins such as BAD, merlin, and FOXOs; and (3) a third group of proteins whose precise role, if any, in malignant transformation is still unknown. These major substrates of PAKs are listed in Table 2.1. For detail of each substrate's biological function, see the next section.

Like many other Ser/Thr kinases, there is some flexibility in the recognition sequences phosphorylated by PAKs. Shown in Figure 2.4 are examples of phosphorylation sites for several PAK substrates. One study used PAK2 and compared a limited number of peptides derived from the substrate KKRKSGL. This yielded a recognition sequence for PAK2 that is characterized by two basic amino acids in the −2 and −3 positions. For example, the peptide (K/R)RXS, in which the −2 position is an Arg and the -3 position is an Arg or a Lys, is efficiently phosphorylated at the Ser residue (X can be an acidic, basic, or neutral amino acid) [138]. A more comprehensive study used a wider array of peptides and found that PAK1 and PAK2 preferred large hydrophobic residues in positions from +1 to +3, in addition to preferring basic amino acids at the −2 and −3 positions [96]. PAK1 and PAK2 have nearly identical substrate specificities, but the substrate specificity of PAK4 is significantly different. PAK4 has a strong preference for Ala at the +2 and Ser at the +3 position. It should be noted that although there are differences in the preferred consensus sequences for Group I (PAK1–3) and Group II PAKs (PAK4–6), most known substrates are phosphorylated by both groups. Additionally, both groups strongly prefer Ser to Thr as a phospho-acceptor site and do not phosphorylate Tyr at all.

PAKs regulate several signaling pathways controlling tumor cell growth, metastasis, and cell survival as well as tumor-induced angiogenesis, including RAF/MEK/ERK, LIM kinase, BAD, merlin, and NFκB (for an outline, see Figure 2.1). In many cases, the direct link between PAKs and their downstream transducers has been

Table 2.1 PAK substrates

Substrate	Sites	Isoform	Reference
Alpha-PIX	S488	Pak1	[96]
Beta-PIX	S340, S525 (transcript A); S497, S682 (transcript B)	Pak1, Pak2	[96,98]
Caldesmon	S657, S687	Pak1, Pak3	[99–101]
CPI17	T38	Pak1	[102]
Filamin A	S2152	Pak1	[103]
GIT1	S517	Pak1	[104]
LIM kinase	T508	PAK1, PAK4	[22,23]
MLCK	S439, S991	Pak1, Pak2	[105,106]
NEF		PAK1	[107]
Stathmin	S16	Pak1	[108]
Rho GDI	S101, S174	Pak1	[109]
R-MLC	S19	Pak2	[110,111]
Vimentin	S25, S38, S50, S56, S65, S72	PAK1	[112–116]
Aurora-A	T288, S342	PAK1	[104]
B-Raf	S446	PAK1	[117]
c-Myc	T358, S373, T400	PAK2	[118]
C-Raf1	S338; S339	PAK1–3, PAK5	[119–125]
β-catenin	S675, S663	PAK1, PAK4	[42,126,127]
ER alpha	S305	PAK1	[54,55]
MEK1	S298	Pak1	[119,128]
MEKK1	S67	Pak1	[129]
Merlin	S518	PAK1	[14]
Prolactin	S179	PAK2	[130]
BAD	S111 (indirectly at S112 and S136)	PAK1, PAK2	[131–135]
DLC1	S88	PAK1	[136]
FOXO	S256	PAK1	[137]

```
Raf-1    (aa 331-445)  :  PRGQRDSSYYWEIE
BAD      (aa 105-119)  :  ETRSRHSSYPAGTE
MEK      (aa 290-304)  :  TPGRPLSSYGMDSR
LIMK     (aa 501-515)  :  DRKKRYTVVGNPYW
Merlin   (aa 511-526)  :  TDMKRLSMEIEKEK
```

Consensus : **(K/R) R** X (S/T) (hydrophobic)

Figure 2.4 PAK phosphorylation sites. PAKs phosphorylate a variety of substrates on Ser/ Thr residues, preferably in the context of basic residues such as K/R, R/X, X, and S/T, for malignant transformation. Shown here are the sequences of phosphorylation sites of several PAK substrates. Consensus sequence is also shown. X can be acidic, basic or neutral amino acids.
Source: Provided by Drs. Diana Ye and Jeff Field.

established. However, in a few cases, such as the activation of the transcription factor NFκB by PAK1/PAK4, the direct targets of PAK1/PAK4 that are responsible for the activation of such oncogenic signal pathways have yet to be determined. The RAF/MEK/ERK, β-catenin, LIM kinase, ER, and integrin-linked kinase (ILK) pathways are among the best-studied examples of PAK effectors regulating the oncogenic pathways, and these will be discussed later in this chapter.

2.8.2 RAF/MEK/ERK Pathways

The canonical MAPK cascade is widely associated with cell proliferation and consists of RAS/RAF1/MEK/ERK. Historically, this was the first cancer-relevant signal shown to be regulated by the PAK pathway. PAKs phosphorylate two mediators of the MAP kinase pathway, MEK1 and RAF1, at Ser 298 and Ser 338, respectively [9,119–122]. While phosphorylation of these sites by PAKs alone is not sufficient to activate RAF1 or MEK1, it significantly facilitates their activation by their upstream activators, RAS/SRC and RAF1, respectively. In short, RAF1 requires three signal transducers, RAS, SRC, and PAK1, for its full (oncogenic) activation. RAS translocates RAF1 from cytoplasm to the plasma membrane, where SRC phosphorylates RAF1 at Tyr 340/341 [139]. For details, see Chapter 3.

2.8.3 LIM Kinase/Cofilin Pathway

There are several established PAK substrates that control cytoskeletal dynamics. The most well-established target is LIM kinase. Both PAK1 and PAK4 phosphorylate LIM kinase at Thr 508 within its activation loop, which stimulates LIM kinase [22,23]. Then LIM kinase phosphorylates cofilin at Ser 3 to inhibit its F-actin severing activity. Thus, by inducing cofilin phosphorylation, PAK1 stimulates actin filaments formation and membrane ruffling, leading to cell migration and cancer invasion/metastasis [22,24] In other words, blocking PAK1 or LIM kinase would be very critical for effective suppression of the so-called "malignancy" (metastasis) of cancer cells, which is the ultimate cause of premature death of cancer patients. For the details of the roles of this pathway in cancer metastasis and neurodegenerative diseases (AD and HD), see Chapters 3 and 6, respectively.

2.8.4 β-Catenin

β-catenin is an oncogenic transcription factor that is active only when it is localized in the nucleus [140]. In normal cells, β-catenin forms a complex with the tumor-suppressor APC and undergoes a rapid degradation in the cytoplasm by ubiquitin-based proteosomes [140]. However, only when RAS is oncogenically mutated (hyperactivated) and APC simultaneously dysfunctions in colon cells, β-catenin is rapidly accumulated in the nucleus, and the malignant transformation of colon cells takes place [141,142]. Interestingly, in the absence of APC, β-catenin is phosphorylated at Ser 675 by PKA [143], and this phosphorylation appears to be essential for its nuclear localization. However, the action of PKA (cAMP-dependent kinase) is RAS independent, and the kinase that is responsible for the RAS-induced phosphorylation of β-catenin still remains to be identified. Interestingly, we and others have found that PAK1 and its effector RAF are essential for the RAS-induced activation of *Cyclin D*1 and *VEGF* genes, respectively, both of which are targets of β-catenin [21,45,51,61], hinting at the possibility that PAK1 may be responsible for this β-catenin phosphorylation as well.

Finally, around 2008 we found that PAK1 forms a complex of β-catenin [144]. Furthermore, in 2012 we demonstrated that siRNA-based silencing of PAK1 blocks both β-catenin phosphorylation and the growth (and metastasis) of human colon (RAS-transformed) cancer cell lines grafted in mice [145]. Independently, a Chinese group also confirmed [42] that downregulation of PAK1 in colon cancer cells reduces total β-catenin level as well as cell proliferation *in vitro*. PAK1 directly phosphorylates β-catenin at Ser 675 leading to more stable and transcriptional active β-catenin [42]. In support of these results, PAK1 is required for full Wnt signaling, and superactivation of β-catenin is achieved by simultaneous knockout of APC and activation of PAK1. Further confirming this conclusion, overexpression of PAK1 is observed in 70% of human colon cancer samples and is correlated with massive β-catenin accumulation [42].

2.8.5   Estrogen Receptor

Around 75% of human breast cancers require estrogen for their growth [62]. Estrogen binds its receptor (ER) in the nucleus to eventually cause the malignant transformation of mammary tissues [62]. Thus, estrogen antagonists such as Tam have been widely used for the therapy of these ER-dependent breast cancers. Interestingly, PAK1 is among the major kinases that directly phosphorylate ER for transactivation [55]. Once the ER is phosphorylated, it becomes estrogen independent and leads to the Tam resistance of these breast cancers [59]. Furthermore, ER activates PAK1 somehow (probably through SRC family kinases) forming a vicious oncogenic cycle of ER and PAK1 [137]. However, we found that potent PAK1 blockers such as FK228 and a combination of two Tyr kinase inhibitors (PP1 and AG 879/GL-2003) almost completely suppress the growth of both Tam-sensitive and -resistant breast cancers as well as the remaining ER-independent breast cancers grafted in mice [32,62].

2.8.6   Integrin-Linked Kinase

ILK, first cloned in 1996 by Shouk Dedhar's group [146], then at the University of Toronto, is a phosphorylated protein that regulates physiological processes that overlap with those regulated by PAK1. Most importantly, both PAK1 and ILK are essential for the growth of majority of solid tumors, but not for normal cell growth [147]. Thus, ILK blockers or PAK1 blockers would be among the almost ideal cancer therapeutics that do not cause any significant side effects. In 2007, Rakesh Kumar's group [148] at the MDACC reported the role of ILK phosphorylation by PAK1 in ILK-mediated signaling and intracellular translocation. They found that PAK1 phosphorylates ILK at both Thr 173 and Ser 246 *in vitro* and *in vivo*. Depletion of

PAK1 decreased the levels of endogenous ILK phosphorylation *in vivo*. Mutation of PAK1 phosphorylation sites on ILK to Ala reduced cell motility and cell proliferation. ILK localizes predominantly in the cytoplasm but also resides in the nucleus. Transfection of MCF-7 cells with point mutants ILK-T173A, ILK-S246A, or a double mutant ILK-T173A/S246A, altered ILK localization. Selective depletion of PAK1 dramatically increased the nuclear and focal point accumulation of ILK, further demonstrating a role for PAK1 in ILK translocation. They also identified functional nuclear localization sequence and nuclear export sequence motifs in ILK. Together, these results suggest that ILK is a PAK1 substrate and undergoes phosphorylation-dependent shuttling between the nucleus and cytoplasm.

Among the first ILK-specific inhibitors is QLT0254, developed in 2005 by David Hedley's group [147] at Princess Margaret Hospital in Toronto, which, administered alone at the dose of 100 mg/kg, i.p., twice a week suppresses by 50% the growth of human pancreatic cancer xenograft in mice and further enhances the sensitivity of this cancer to gemcitabine (80 mg/kg, i.p., twice a week). However, it should be worth to note that unlike PAK1 blockers, ILK inhibitors do not directly block the oncogenic RAF/MEK/ERK and LIM kinase/cofilin signal pathways, and therefore ILK blockers in general are significantly less effective than PAK1 blockers to suppress the growth of PAK1/ILK-dependent solid tumors.

More recently Ching Shih Chen's group [149] at OSU identified a novel, more potent ILK inhibitor called Compound 22 (or OSU-22): *N*-methyl-3-(1-(4-(piperazin-1-yl) phenyl)-5-(4′-(trifluoromethyl)-[1,1′-biphenyl]-4-yl)-1*H*-pyrazol-3-yl) propanamide (see Figure 2.5). Its IC_{50} is around 0.6 μM, and it exhibited high *in vitro* potency against a panel of prostate and breast cancer cell lines with IC_{50}, 1–2.5 μM, with no effect on normal epithelial cell growth. OSU-22 leads to the dephosphorylation of AKT at Ser-473 and other ILK targets, including GSK-3β and the myosin light chain. Moreover, OSU-22 suppresses the expression of the transcription/translation factor YB-1 and its targets ErbB1 and ErbB2 in PC-3 (human prostate cancer) cells, which could be rescued by the CA mutant of ILK. Together, this broad spectrum of mechanisms underlies the therapeutic potential of OSU-22 in cancer treatment. However, OSU-22 alone (50 mg/kg, p.o., daily) suppresses the PAK1-dependent growth of PC-3 cancer xenograft in mice by only 50% [149], suggesting again that for complete growth suppression (by blocking the remaining

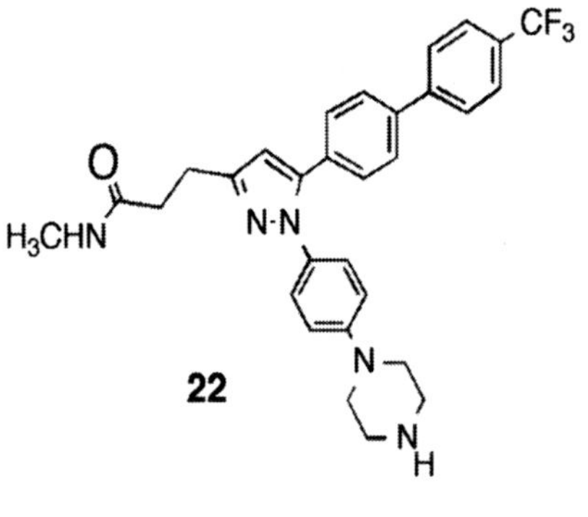

Figure 2.5 ILK inhibitor Compound 22 (OSU-22).

effectors of PAK1), PAK1 blockers such as FK228 and propolis would be far more effective than this ILK inhibitor alone, which does not fully inhibit PAK1.

2.8.7   c-Jun N-Terminal Kinase

c-Jun is an oncogenic transcription factor that is activated by RAS or UV irradiation [150]. Mike Karin's group [150] at UCSD found in 1993 that the N-terminal domain of c-Jun is phosphorylated at Ser residues 63 and 73 by an oncogenic kinase called c-Jun N-terminal kinase (JNK) for transactivation. However, the kinase(s) responsible for the RAS-induced activation of JNK remained to be identified.

Richard Cerione's group [151] at Cornell University revealed in 1995 that a CA mutant of PAK3 could activate JNK. Subsequently, the CA mutant of PAK1 (L107F) was also shown to activate JNK [152]. In support of the PAK1/PAK3–mediated JNK activation, we found that the compound CEP-1347, which inhibits both PAK1 and PAK3 directly, blocks JNK activation [153,154]. Later PAK5 was also found to activate JNK [155]. Furthermore, overexpression of the DN mutant of PAK4 reduces JNK activation, indicating that PAK4 is also required in part for the activation of JNK [156]. These observations established that a variety of PAKs could mediate the RAS/UV-induced activation of JNK. Nevertheless, it still remained to be clarified whether these PAKs directly phosphorylate JNK or need another downstream kinase (effector) for JNK activation.

2.8.8   Aurora-A

Ed Manser's group [104] in Singapore found that PAK1 in the complex of PIX and GIT phosphorylates an oncogenic kinase called Aurora-A at Thr288 and Ser 342 for activation. Aurora-A is associated with microtubule (MT)-based centrosomes and is responsible for the maturation of centrosomes during mitosis. Blocking either PAK1 or PIX significantly delays maturation of centrosomes [104].

2.8.9   BAD/Bcl-2 Pathway

Apoptosis, or programmed cell death, is a fundamental process in the development of multicellular organisms. Apoptosis enables an organism to eliminate unwanted or defective cells through an organized process of cellular disintegration. It is a prominent tumor-suppression process, and cancer cells require inactivation of proapoptotic pathways for tumor formation and progression. PAKs have been shown to downregulate several important proapoptotic pathways.

In 2000, Gary Bokoch's group [131] at Scripps in San Diego reported that PAK1 phosphorylates a proapoptotic protein called BAD at Ser 112 to inactivate its ability to bind/antagonize the antiapoptotic Bcl-2, leading to the protection of cancer cells from apoptosis. However, recent studies by Jeff Field's group (and others) [132] suggest that although PAK1 directly phosphorylates BAD at Ser 111 *in vitro*, the kinase that directly phosphorylates BAD at Ser 112 *in vivo* (cells) is RAF1, an effector of PAK1, and not PAK1 *per se*.

In other words, it is more likely that PAK1 protects cells from their apoptosis via the RAF1–BAD pathway. PAK1 and PAK5 phosphorylate RAF1 at Ser 338, leading to the translocation of RAF1 to the mitochondria [122,132,133,157]. At the mitochondria, RAF1 forms a complex with Bcl-2 and phosphorylates BAD at Ser 112 [132]. Bcl-2 is a protooncogene that maintains the integrity of the mitochondrial barrier if bound in protective complexes, whereas binding of Bcl-2 to BAD induces release of proapoptotic factors from the mitochondria, leading to apoptosis. Phosphorylation of BAD at specific sites, including Ser 112, renders the cells to apoptosis. In support of this notion, RAF1-KO (knockout) cells have higher rates of apoptosis than the control cells [158].

2.8.10   FOXO (FKHR)

FOXO (FKHR) family transcription factors are tumor suppressors that are activated by AMPK and inactivated by PAK1 through specific phosphorylation [137,159]. FOXO is essential for longevity and required for activation of a variety of genes including *Hsp*16, encoding a small heat shock protein that serves as an ATP-dependent chaperone [36]. Thus, PAK1 shortens the lifespan of *Caenorhabditis elegans* and makes this worm heat sensitive by inactivating FOXO [36], as discussed in detail in Chapter 7. In other words, any PAK1 blockers or AMPK activators (often these compounds, such as curcumin, do both jobs) would be useful not only for cancer therapy but also for extending the healthy lifespan significantly [36].

2.8.11   Merlin

The *NF*2 gene product "merlin" is a direct PAK1 inhibitor of 595 amino acids [74]. Thus, in NF2-deficient tumors such as meningioma and schwannoma as well as around a half of mesothelioma (HMM), PAK1 is hyperactivated, and their growth requires PAK1 [74]. Interestingly, however, merlin is phosphorylated by PAK1 or PKA at Ser 518 and loses its tumor-suppressing activity [14,15,160]. How does this phosphorylation inactivate the tumor-suppressive function of merlin? It is conceivable that phosphorylation-induced heterodimerization of merlin with its structurally related protein, ezrin, blocks the interaction of merlin with the cell surface receptor CD44 [85,160]. CD44 is an oncogenic receptor of hyaluronic acid (HA), and the merlin–CD44 interaction, which blocks the oncogenic HA–CD44 interaction, is essential for contact inhibition of both growth and movement of normal cells [85,160].

2.9   Concluding Remarks

Here we have discussed how PAKs, in particular PAK1 and PAK4, promote the anchorage-independent growth, angiogenesis, and metastasis of major solid tumors by regulating either their major oncogenic or tumor-suppressive substrates, emphasizing the point that these solid (malignant or benign) tumors will be major potential chemotherapeutic targets of anti-PAK1 or -PAK4 drugs in the future. As discussed in Chapter 3, more than a dozen PAK1 blockers (synthetic or natural), such as FK228 and propolis, and a few mostly synthetic PAK4 blockers such as PF-3758309 have been developed or identified during the last decade, and it is hoped that they will replace conventional chemotherapies (DNA/RNA/MT poisons), which cause a variety of side effects, in the not-too-distant future.

Although Gleevec/STI-571, an inhibitor of Tyr kinases (ABL/c-KIT/PDGFR), was successfully developed around the turn of this century as the very first "signal therapeutic" (antioncogenic signal blocker) and serves as a sort of "miracle" drug for a few rare cancers such as chronic myelogenous leukemia and gastrointestinal stromal tumor [161,162], these cancers altogether represent less than 0.1 % of all human cancers, and Gleevec is basically useless for the remaining more than 99% of human cancers whose growth is independent of these specific Tyr kinases.

As you will see in the following several chapters, PAKs, in particular PAK1, cause not only a variety of cancers but also several noncancerous diseases or disorders such as type 2 diabetes, hypertension, inflammatory diseases (asthma and arthritis), neurodegenerative diseases (AD and HD), several other brain diseases (TSC, epilepsy, depression, schizophrenia, LD, and autism), and several infectious diseases (malaria, AIDS, and flu) and are clearly responsible for shortening our healthy lifespan. Thus, these PAK1/PAK4 blockers would potentially make a great contribution to the improvement in the coming centuries of our quality of life by alleviating conditions caused by the majority of cancers and a variety of other PAK1-dependent diseases and disorders.

Acknowledgments

We are grateful to Drs. Diana Ye and Jeff Field for their kind suggestions and useful information (including Figure 2.4), which helped us to draft this chapter.

References

[1]   Hamilton SR. Molecular genetics of colorectal carcinoma. Cancer 1992;70:1216–21.
[2]   Bos JL. Ras oncogenes in human cancer: a review. Cancer Res 1989;49:4682–9.
[3]   Gibbs JB, Oliff A, Kohl NE. Farnesyltransferase inhibitors: Ras research yields a potential cancer therapeutic. Cell 1994;77:175–8.
[4]   Downward J. Ras signalling and apoptosis. Curr Opin Genet Dev 1998;8:49–54.
[5]   Qiu RG, Chen J, Kirn D, McCormick F, Symons M. An essential role for Rac in Ras transformation. Nature 1995;374:457–9.
[6]   Qiu RG, Abo A, McCormick F, Symons M. Cdc42 regulates anchorage-independent growth and is necessary for Ras transformation. Mol Cell Biol 1997;17:3449–58.
[7]   Manser E, Leung T, Salihuddin H, Zhao ZS, Lim L. A brain serine/threonine protein kinase activated by Cdc42 and Rac1. Nature 1994;367:40–6.
[8]   Manser E, Leung T, Salihuddin H, Tan L, Lim L. A non-receptor tyrosine kinase that inhibits the GTPase activity of p21cdc42. Nature 1993;363:364–7.

[9] Tang Y, Chen Z, Ambrose D, Liu J, Gibbs J, Chernoff J, et al. Kinase-deficient Pak1 mutants inhibit Ras transformation of Rat-1 fibroblasts. Mol Cell Biol 1997;17:4454–64.

[10] Tang Y, Marwaha S, Rutkowski J, Tennekoon G, Phillips P, Field J. A role for Pak protein kinases in Schwann cell transformation. Proc Natl Acad Sci USA 1998;95:5139–44.

[11] Nur-E-Kamal MSA, Kamal JM, Quresh MM, Maruta H. The CDC42-specific inhibitor derived from ACK blocks the v-Ha-RAS-induced transformation. Oncogene 1999;18:7787–93.

[12] Abercrombie M. Contact inhibition: the phenomenon and its biological implications. Natl Cancer Inst Monogr 1967;26:249–77.

[13] Nwariaku F, Liu Z, Zhu X, Nahari D, Ingle C, Wu RF, et al. NADPH oxidase mediates vascular endothelial cadherin phosphorylation and endothelial dysfunction. Blood 2004;104:3214–20.

[14] Xiao GH, Beeser A, Chernoff J, Testa JR. *PAK links Rac/Cdc42 signaling to Merlin.* J Biol Chem 2002;277:883–6.

[15] Alfthan K, Heiska L, Grönholm M, Renkema G, Carpén. O. PKA phosphorylates merlin at serine 518 independently of PAK and promotes merlin–ezrin hetero-dimerization. J Biol Chem 2004;279:18559–18566.

[16] He H, Levitzki A, Zhu HJ, Walker F, Burgess A, Maruta H. Platelet-derived growth factor requires epidermal growth factor receptor to activate p21-activated kinase family kinases. J Biol Chem 2001;276:26741–26744.

[17] Cifone MA. In vitro growth characteristics associated with benign and metastatic variants of tumor cells. Cancer Metastasis Rev 1982;1:335–47.

[18] Yang S, Kim HM. The RhoA–ROCK–PTEN pathway as a molecular switch for anchorage dependent cell behavior. Biomaterials 2012;33:2902–15.

[19] del Pozo M, Price L, Alderson N, Ren XD, Schwartz M. Adhesion to the extra-cellular matrix regulates the coupling of the small GTPase Rac to its effector PAK. EMBO J 2000;19:2008–14.

[20] Knowles H, Phillips R. Identification of differentially expressed genes in experimental models of the tumor microenvironment using differential display. Anticancer Res 2001;21:2305–11.

[21] Bagheri-Yarmand R, Vadlamudi R, Wang R, Mendelsohn J, Kumar R. VEGF up-regulation via PAK1 signaling regulates heregulin-beta1-mediated angiogenesis. J Biol Chem 2000;275:39451–39457.

[22] Edwards D, Sanders L, Bokoch G, Gill G. Activation of LIM-kinase by Pak1 couples Rac/Cdc42 GTPase signalling to actin cytoskeletal dynamics. Nat Cell Biol 1999;1:253–9.

[23] Dan C, Kelly A, Bernard O, Minden A. Cytoskeletal changes regulated by the PAK4 serine/threonine kinase are mediated by LIM kinase 1 and cofilin. J Biol Chem 2001;276:32115–32121.

[24] Yoshioka K, Foletta V, Bernard O, Itoh K. A role for LIM kinase in cancer invasion. Proc Natl Acad Sci USA 2003;100:7247–52.

[25] Parsons D, Wang TL, Samuels Y, Bardelli A, Cummins J, DeLong L, et al. Colorectal cancer: mutations in a signalling pathway. Nature 2005;436:792.

[26] Whale A, Dart A, Holt M, Jones G, Wells C. PAK4 kinase activity and somatic mutation promote carcinoma cell motility and influence inhibitor sensitivity. Oncogene 2012 (in press)

[27] Callow M, Clairvoyant F, Zhu S, Schryver B, Whyte D, Bischoff J, et al. Requirement for PAK4 in the anchorage-independent growth of human cancer cell lines. J Biol Chem 2002;277:550–8.

[28] Brown L, Kalloger S, Miller M, Shih Ie, M, McKinney S, Santos J, et al. Amplification of 11q13 in ovarian carcinoma. Genes Chromosomes Cancer 2008;47:481–9.

[29] Bostner J, Ahnstrom Waltersson M, Fornander T, Skoog L, Nordenskjold B, Stal O. Amplification of CCND1 and PAK1 as predictors of recurrence and tamoxifen resistance in postmenopausal breast cancer. Oncogene 2007;26:6997–7005.

[30] Ong CC, Jubb A, Haverty P, Zhou W, Tran V, Truong T, et al. Targeting PAK1 to induce apoptosis of tumor cells. Proc Natl Acad Sci USA 2011;108:7177–82.

[31] Chen S, Auletta T, Dovirak O, Hutter C, Kuntz K, El-Ftesi S, et al. Copy number alterations in pancreatic cancer identify recurrent PAK4 amplification. Cancer Biol Ther 2008;7:1793–802.

[32] Reddy S, Ohshiro K, Rayala S, Kumar R. MicroRNA-7, a homeobox D10 target, inhibits PAK 1 and regulates its functions. Cancer Res 2008;68:8195–200.

[33] Hirokawa Y, Levitzki A, Lessene G, Baell J, et al. Signal therapy of human pancreatic cancer and NF1-deficient breast cancer xenograft in mice by a combination of PP1 and GL-2003, anti-PAK1 Drugs (Tyr-kinase inhibitors). Cancer Lett 2007;245:242–51.

[34] Demestre M, Messerli S, Celli N, Shahhossini M, et al. CAPE (caffeic acid phenethyl ester)-based propolis extract (Bio 30) suppresses the growth of human neurofibromatosis (NF) tumor xenografts in mice. Phytother Res 2009;23:226–30.

[35] Maruta H. An innovated approach to in vivo screening for the major anti-cancer drugs Horizons in cancer research, vol. 41. : Nova Science Publishers; 2010. pp. 249–59.

[36] Maruta H. Effective neurofibromatosis therapeutics blocking the oncogenic kinase PAK1. Drug Discov Ther 2011;5:266–78.

[37] Ying JE, Zhu LM, Liu BX. Developments in metastatic pancreatic cancer: is gemcitabine still the standard?. World J Gastroenterol 2012;18:736–45.

[38] Kinzler K, Vogelstein B. The colorectal cancer gene hunt: current findings. Hosp Pract (Off Ed) 1992;27:51–8.

[39] D'Abaco GM, Whitehead RH, Burgess AW. Synergy between Apc min and an activated ras mutation is sufficient to induce colon carcinomas. Mol Cell Biol 1996;16:884–91.

[40] Hinds P, Finlay C, Levine A. Mutation is required to activate the p53 gene for cooperation with the ras oncogene and transformation. J Virol 1989;63:739–46.

[41] Easwaran V, Song V, Polakis P, Byers S. The ubiquitin-proteasome pathway and serine kinase activity modulate adenomatous polyposis coli protein-mediated regulation of beta-catenin-lymphocyte enhancer-binding factor signaling. J Biol Chem 1999;274:16641–16645.

[42] Zhu G, Wang Y, Huang B, Liang L, Ding Y, Xu A, et al. A Rac1/PAK1 cascade controls beta-catenin activation in colon cancer cells. Oncogene 2012;31:1001–12.

[43] Huynh N, Liu KH, Baldwin G, He H. PAK 1 stimulates colon cancer cell growth and migration/invasion via ERK- and AKT-dependent pathways. Biochim Biophys Acta 2010;1803:1106–13.

[44] He H, Huynh N, Liu K, Malcontenti-Wilson C, et al. PAK1 knockdown inhibits β-catenin signalling and blocks colorectal cancer growth. Cancer Lett 2012;317:65–71.

[45] Tetsu O, McCormick F. Beta-catenin regulates expression of cyclin D1 in colon carcinoma cells. Nature 1999;398:422–6.

[46] Grazia Lampugnani M, Zanetti A, Corada M, Takahashi T, Balconi G, Breviario F, et al. Contact inhibition of VEGF-induced proliferation requires vascular endothelial cadherin, beta-catenin, and the phosphatase DEP-1/CD148. J Cell Biol 2003;161:793–804.

[47] Higashi H, Tsutsumi R, Fujita A, Yamazaki S, Asaka M, Azuma T, et al. Biological activity of the *Helicobacter pylori* virulence factor CagA is determined by variation in the tyrosine phosphorylation sites. Proc Natl Acad Sci USA 2002;99:14428–14433.

[48] Baek HY, Lim JW, Kim H. Interaction between the Helicobacter pylori CagA and alpha-Pix in gastric epithelial AGS cells. Ann NY Acad Sci 2007;1096:18–23.

[49] Murray B, Guo CX, Piraino J, Westwick J, et al. Small-molecule PAK inhibitor PF-3758309 is a potent inhibitor of oncogenic signaling and tumor growth. Proc Natl Acad Sci USA 2010;107:9446–51.

[50] Balasenthil S, Sahin A, Barnes C, Wang R, Pestell R, Vadlamudi R, et al. PAK1 signaling mediates cyclin D1 expression in mammary epithelial and cancer cells. J Biol Chem 2004;279:1422–8.

[51] Wang RA, Zhang H, Balasenthil S, Medina D, Kumar R. PAK1 hyperactivation is sufficient for mammary gland tumor formation. Oncogene 2006;25:2931–6.

[52] Liu Y, Chen N, Cui X, Zheng X, Deng L, Price S, et al. Pak4 disrupts mammary acinar architecture and promotes mammary tumorigenesis. Oncogene 2010;29:5883–94.

[53] Wang RA, Mazumdar A, Vadlamudi R, Kumar R. PAK1 phosphorylates and transactivates estrogen receptor-alpha and promotes hyperplasia in mammary epithelium. EMBO J 2002;21:5437–47.

[54] Rayala S, Talukder A, Balasenthil S, Tharakan R, Barnes C, Wang RA, et al. PAK 1 regulation of estrogen receptor-alpha activation involves serine 305 activation linked with serine 118 phosphorylation. Cancer Res 2006;66:694–701.

[55] Shrestha Y, Schafer E, Boehm J, Thomas S, He F, Du J, et al. PAK1 is a breast cancer oncogene that coordinately activates MAPK and MET signaling. Oncogene 2012;31:3397–408.

[56] Du J, Sun C, Hu Z, Yang Y, Zhu Y, Zheng D, et al. Lysophosphatidic acid induces MDA-MB-231 breast cancer cells migration through activation of PI3K/PAK1/ERK signaling. PLoS One 2010;5:e15940.

[57] Friedland J, Lakins J, Kazanietz M, Chernoff J, Boettiger D, Weaver V. α6β4 integrin activates Rac-dependent PAK1 to drive NF-κB-dependent resistance to apoptosis in 3D mammary acini. J Cell Sci 2007;120:3700–12.

[58] Bostner J, Skoog L, Fornander T, Nordenskjold B, Stal O. Estrogen receptor-alpha phosphorylation at serine 305, nuclear PAK 1 expression, and response to tamoxifen in postmenopausal breast cancer. Clin Cancer Res 2010;16:1624–33.

[59] Nheu T, He H, Hirokawa Y, Walker F, Wood J, Maruta H. PAK is essential for RAS-induced upregulation of cyclin D1 during the G1 to S transition. Cell Cycle 2004;3:71–4.

[60] Hirokawa Y, Arnold M, Nakajima H, Zalcberg J, et al. Signal therapy of breast cancer xenograft in mice by the HDAC inhibitor FK228 that blocks the activation of PAK1 and abrogates the tamoxifen-resistance. Cancer Biol Ther 2005;4:956–60.

[61] Aksamitiene E, Achanta S, Kolch W, Kholodenko B, Hoek J, Kiyatkin A. Prolactin-stimulated activation of ERK1/2 mitogen-activated protein kinases is controlled by PI-3-kinase/Rac/PAK signaling pathway in breast cancer cells. Cell Signal 2011;23:1794–805.

[62] Tao J, Oladimeji P, Rider L, Diakonova M. PAK1-Nck regulates cyclin D1 promoter activity in response to prolactin. Mol Endocrinol 2011;25:1565–78.

[63] Shalaby M, Shepard H, Presta L, Rodrigues M, Beverley P, Feldmann M, et al. Development of humanized bispecific antibodies reactive with cytotoxic lymphocytes and tumor cells overexpressing the HER2 protooncogene. J Exp Med 1992;175:217–25.

[64] Arias-Romero L, Villamar-Cruz O, Pacheco A, Kosoff R, Huang M, Muthuswamy S, et al. A Rac–Pak signaling pathway is essential for ErbB2-mediated transformation of human breast epithelial cancer cells. Oncogene 2010;29:5839–49.

[65] Pickl M, Ries C. Comparison of 3D and 2D tumor models reveals enhanced HER2 activation in 3D associated with an increased response to trastuzumab. Oncogene 2009;28:461–8.

[66] zur Hausen H. Papillomaviruses in anogenital cancer as a model to understand the role of viruses in human cancers. Cancer Res 1989;49:4677–81.

[67] Chan JK, Berek JS. Impact of the human papilloma vaccine on cervical cancer. J Clin Oncol 2007;25:2975–82.

[68] Wu R, Abramson A, Symons M, Steinberg B. Pak1 and Pak2 are activated in recurrent respiratory papillomas, contributing to one pathway of Rac1-mediated COX-2 expression. Int J Cancer 2010;127:2230–7.

[69] Singh M, Singh N. Curcumin counteracts the proliferative effect of estradiol and induces apoptosis in cervical cancer cells. Mol Cell Biochem 2011;347:1–11.

[70] Li J, Yen C, Liaw D, Podsypanina K, et al. PTEN, a putative protein tyrosine phosphatase gene mutated in human brain, breast, and prostate cancer. Science 1997:1943–7.

[71] Sasakawa Y, Naoe Y, Noto T, Inoue T, Sasakawa T, Matsuo M, et al. Antitumor efficacy of FK228, a novel histone deacetylase inhibitor, depends on the effect on expression of angiogenesis factors. Biochem Pharmacol 2003;66:897–906.

[72] Suzuki H, Freije D, Nusskern D, Okami K, Cairns P, Sidransky D, et al. Interfocal heterogeneity of PTEN/MMAC1 gene alterations in multiple metastatic prostate cancer tissues. Cancer Res 1998;58:204–9.

[73] Sundram V, Chauhan S, Ebeling M, Jaggi M. Curcumin attenuates β-catenin signaling in prostate cancer cells through activation of protein kinase D1. PLoS One 2012;7:e35368.

[74] Hirokawa Y, Tikoo A, Huynh J, Utermark T, et al. A clue to the therapy of neurofibromatosis type 2: NF2/merlin is a PAK1 inhibitor. Cancer J 2004;10:20–6.

[75] Rodenhuis S, Slebos RJ. The ras oncogenes in human lung cancer. Am Rev Respir Dis 1990;142:S27–30.

[76] Kissil J, Walmsley M, Hanlon L, Haigis K, Bender Kim C, Sweet-Cordero A, et al. Requirement for Rac1 in a K-ras induced lung cancer in the mouse. Cancer Res 2007;67:8089–94.

[77] Brustugun O, Helland A, Fjellbirkeland L, Kleinberg L, Ariansen S, Jebsen P, et al. Mutation testing for non-small-cell lung cancer. Tidsskr Nor Laegeforen 2012;132:952–5.

[78] Bianchi A, Mitsunaga S, Cheng JQ, Klein W, Jhanwar S, Seizinger B, et al. High frequency of inactivating mutations in the neurofibromatosis type 2 gene (NF2) in primary malignant mesotheliomas. Proc Natl Acad Sci USA 1995;92:10854–10858.

[79] Schraml P, Schwerdtfeger G, Burkhalter F, Raggi A, et al. Combined array comparative genomic hybridization and tissue microarray analysis suggest *PAK1* at 11q13.5-q14 as a critical oncogene target in ovarian carcinoma. Am J Pathol 2003;163:985–92.

[80] Hashimoto H, Sudo T, Maruta H, Nishimura R. The direct PAK1 inhibitor, TAT-PAK18, blocks preferentially the growth of human ovarian cancer cell lines in which PAK1 is abnormally activated by autophosphorylation at Thr 423. Drug Discov Ther 2010;4:1–4.

[81] Hashimoto H, Messerli S, Sudo T, Maruta H. Ivermectin inactivates the kinase PAK1 and block the PAK1-dependent growth of human ovarian cancer and NF2 tumor cell lines. Drug Discov Ther 2009;3:243–6.

[82] Porchia L, Guerra M, Wang YC, Zhang YL, et al. OSU-03012, a celecoxib derivative, directly targets PAK. Mol Pharmacol 2007;72:1124–31.

[83] Zhu J, Huang JW, Tseng PH, Yang YT, Fowble J, Shiau CW, et al. From the cyclooxygenase-2 inhibitor celecoxib to a novel class of 3-phosphoinositide-dependent protein kinase-1 inhibitors. Cancer Res 2004;64:4309–18.

[84] Lee TX, Packer M, Huang J, Akhmametyeva E, Kulp S, Chen CS, et al. Growth inhibitory and anti-tumour activities of OSU-03012, a novel PDK-1 inhibitor, on vestibular schwannoma and malignant schwannoma cells. Eur J Cancer 2009;45:1709–20.

[85] Stamenkovic I, Yu Q. Merlin, a "magic" linker between extracellular cues and intracellular signaling pathways that regulate cell motility, proliferation, and survival. Curr Protein Pept Sci 2010;11:471–84.

[86] Bai LY, Weng JR, Tsai CH, Sargeant A, et al. OSU-03012 sensitizes TIB-196 myeloma cells to imanitib mesylate via AMPK and STAT3 pathways. Leuk Res 2010;34:826–30.

[87] Khan SB, Maududi T, Barton K, Ayers J, Alkan S. Analysis of histone deacetylase inhibitor, depsipeptide (FR901228), effect on multiple myeloma. Br J Haematol 2004;125:156–61.

[88] Bharti A, Donato N, Singh S, Aggarwal B. Curcumin (diferuloylmethane) down-regulates the constitutive activation of nuclear factor-kappa B and IkappaBalpha kinase in human multiple myeloma cells, leading to suppression of proliferation and induction of apoptosis. Blood 2003;101:1053–62.

[89] Sawa H, Murakami H, Kumagai M, Nakasako M, et al. HDAC inhibitor, FK228, induces apoptosis and suppresses cell proliferation of human glioblastoma cells in vitro and in vivo. Acta Neuropathol 2004;107:523–31.

[90] Kuo HC, Kuo WH, Lee YJ, Lee WL, et al. Bottom of form inhibitory effect of caffeic acid phenethyl ester on the growth of C6 glioma cells in vitro and in vivo. Cancer Lett 2006;234:199–208.

[91] Aoki H, Takada Y, Kondo S, Sawaya R, et al. Evidence that curcumin suppresses the growth of malignant gliomas in vitro and in vivo through induction of autophagy: role of AKT and ERK signaling pathways. Mol Pharmacol 2007;72:29–39.

[92] Ching YP, Leong VY, Lee MF, Xu HT, Jin DY, Ng. IO. PAK is overexpressed in hepatocellular carcinoma and enhances cancer metastasis involving JNK activation and paxillin phosphorylation. Cancer Res 2007;67:3601–8.

[93] Darvesh A, Aggarwal B, Bishayee A. Curcumin and liver cancer: a review. Curr Pharm Biotechnol 2012;13:218–28.

[94] Pavey S, Zuidervaart W, van Nieuwpoort F, Packer L, Jager M, Gruis N, et al. Increased PAK-1 expression is associated with invasive potential in uveal melanoma. Melanoma Res 2006;16:285–96.

[95] Drinyaev V, Mosin V, Kluglyak E, Novik. T, et al. Antitumor effect of avermectin. Eur J Pharmacol 2004;501:19–23.

[96] Rennefahrt U, Deacon S, Parker S, Devarajan K, Beeser A, Chernoff J, et al. Specificity profiling of pak kinases allows identification of novel phosphorylation site. J Biol Chem 2007;282:15667–15678.

[97] Dummler B, Ohshiro K, Kumar R, Field J. Pak protein kinases and their role in cancer. Cancer Metastasis Rev 2009;28:51–63.

[98] Shin EY, Shin KS, Lee CS, Woo KN, Quan SH, Soung NK, et al. Phosphorylation of p85 beta PIX, a Rac/Cdc42-specific guanine nucleotide exchange factor, via the Ras/ERK/PAK2 pathway is required for basic fibroblast growth factor-induced neurite outgrowth. J Biol Chem 2002;277:44417–44430.

[99]   Foster D, Shen LH, Kelly J, Thibault P, Van Eyk J, Mak AS. Phosphorylation of caldesmon by PAK. Implications for the Ca(2+) sensitivity of smooth muscle contraction. J Biol Chem 2000;275:1959–65.

[100] McFawn P, Shen L, Vincent S, Mak A, Van Eyk J, Fisher J. Calcium-independent contraction and sensitization of airway smooth muscle by PAK. Am J Physiol Lung Cell Mol Physiol 2003;284:L863–70.

[101] Van Eyk J, Arrell D, Foster D, Strauss J, Heinonen T, Furmaniak-Kazmierczak E, et al. Different molecular mechanisms for Rho family GTPase-dependent, Ca^{2+}-independent contraction of smooth muscle. J Biol Chem 1998;273:23433–23439.

[102] Takizawa N, Koga Y, Ikebe M. Phosphorylation of CPI17 and myosin binding subunit of type 1 protein phosphatase by PAK. Biochem Biophys Res Commun 2002;297:773–8.

[103] Vadlamudi R, Li F, Adam L, Nguyen D, Ohta Y, Stossel T, et al. Filamin is essential in actin cytoskeletal assembly mediated by PAK 1. Nat Cell Biol 2002;4:681–90.

[104] Zhao ZS, Lim JP, Ng YW, Lim L, Manser E. The GIT-associated kinase PAK targets to the centrosome and regulates Aurora-A. Mol Cell 2005;20:237–49.

[105] Sanders L, Matsumura F, Bokoch G, de Lanerolle P. Inhibition of myosin light chain kinase by PAK. Science 1999;283:2083–5.

[106] Goeckeler Z, Masaracchia R, Zeng Q, Chew TL, Gallagher P, Wysolmerski R. Phosphorylation of myosin light chain kinase by PAK2. J Biol Chem 2000;275: 18366–18374.

[107] Nguyen DG, Wolff K, Yin H, Caldwell J, et al. "UnPAKing" human immuno-deficiency virus (HIV) replication: using small interfering RNA screening to identify novel cofactors and elucidate the role of group I PAKs in HIV infection. J Virol 2006;80:130–7.

[108] Daub H, Gevaert K, Vandekerckhove J, Sobel A, Hall A. Rac/Cdc42 and p65PAK regulate the microtubule-destabilizing protein stathmin through phosphorylation at serine 16. J Biol Chem 2001;276:1677–80.

[109] DerMardirossian C, Schnelzer A, Bokoch G. Phosphorylation of RhoGDI by Pak1 mediates dissociation of Rac GTPase. Mol Cell 2004;15:117–27.

[110] Chew TL, Masaracchia R, Goeckeler Z, Wysolmerski R. Phosphorylation of non-muscle myosin II regulatory light chain by p21-activated kinase (gamma-PAK). J Muscle Res Cell Motil 1998;19:839–54.

[111] Ramos E, Wysolmerski R, Masaracchia R. Myosin phosphorylation by human cdc42-dependent S6/H4 kinase/gammaPAK from placenta and lymphoid cells. Recept Signal Transduct 1997;7:99–110.

[112] Goto H, Tanabe K, Manser E, Lim L, Yasui Y, Inagaki M. Phosphorylation and reorganization of vimentin by PAK. Genes Cells 2002;7:91–7.

[113] Li QF, Spinelli A, Wang R, Anfinogenova Y, Singer H, Tang DD. Critical role of vimentin phosphorylation at Ser-56 by p21-activated kinase in vimentin cytoskeleton signaling. J Biol Chem 2006;281:34716–34724.

[114] Tang DD, Bai Y, Gunst SJ. Silencing of PAK attenuates vimentin phosphorylation on Ser-56 and reorientation of the vimentin network during stimulation of smooth muscle cells by 5-hydroxytryptamine. Biochem J 2005;388:773–83.

[115] Wang R, Li QF, Anfinogenova Y, Tang DD. Dissociation of Crk-associated substrate from the vimentin network is regulated by PAK on ACh activation of airway smooth muscle. Am J Physiol Lung Cell Mol Physiol 2007;292:L240–8.

[116] Chan W, Kozma R, Yasui Y, Inagaki M, Leung T, Manser E, et al. Vimentin intermediate filament reorganization by Cdc42: involvement of PAK and p70 S6 kinase. Eur J Cell Biol 2002;81:692–701.

[117] Tran NH, Wu X, Frost J. B-Raf and Raf-1 are regulated by distinct autoregulatory mechanisms. J Biol Chem 2005;280:16244–16253.

[118] Huang Z, Traugh J, Bishop J. Negative control of the Myc protein by the stress-responsive kinase Pak2. Mol Cell Biol 2004;24:1582–94.

[119] Beeser A, Jaffer Z, Hofmann C, Chernoff J. Role of group A p21-activated kinases in activation of extracellular-regulated kinase by growth factors. J Biol Chem 2005;280:36609–36615.

[120] Tran NH, Frost J. Phosphorylation of Raf-1 by p21-activated kinase 1 and Src regulates Raf-1 autoinhibition. J Biol Chem 2003;278:11221–11226.

[121] King AJ, Sun H, Diaz B, Barnard D, Miao W, Bagrodia S, et al. The protein kinase Pak3 positively regulates Raf-1 activity through phosphorylation of serine 338. Nature 1998;396:180–3.

[122] Wu X, Carr H, Dan I, Ruvolo P, Frost J. PAK 5 activates Raf-1 and targets it to mitochondria. J Cell Biochem 2008;105:167–75.

[123] Edin M, Juliano R. Raf-1 serine 338 phosphorylation plays a key role in adhesion-dependent activation of extracellular signal-regulated kinase by epidermal growth factor. Mol Cell Biol 2005;25:4466–75.

[124] Chaudhary A, King W, Mattaliano M, Frost J, Diaz B, et al. Phosphatidylinositol 3-kinase regulates Raf1 through Pak phosphorylation of serine 338. Curr Biol 2000;10:551–4.

[125] Zang M, Hayne C, Luo Z. Interaction between active Pak1 and Raf-1 is necessary for phosphorylation and activation of Raf-1. J Biol Chem 2002;277:4395–405.

[126] Li Y, Shao Y, Tong Y, Shen T, et al. Nucleo-cytoplasmic shuttling of PAK4 modulates β-catenin intracellular translocation and signaling. Biochim Biophys Acta 2012;1823:465–75.

[127] Park MH, Kim DJ, You ST, Lee CS, Kim HK, Park SM, et al. Phosphorylation of β-catenin at serine 663 regulates its transcriptional activity. Biochem Biophys Res Commun 2012 (in press).

[128] Frost JA, Xu S, Hutchison M, Marcus S, Cobb M. Actions of Rho family small G proteins and PAKs on mitogen-activated protein kinase family members. Mol Cell Biol 1996;16:3707–13.

[129] Gallagher E, Xu S, Moomaw C, Slaughter C, Cobb M. Binding of JNK/SAPK to MEKK1 is regulated by phosphorylation. J Biol Chem 2002;277:45785–45792.

[130] Tuazon P, Lorenson M, Walker A, Traugh JA. p21-activated protein kinase gamma-PAK in pituitary secretory granules phosphorylates prolactin. FEBS Lett 2002;515:84–8.

[131] Schurmann A, Mooney A, Sanders L, Sells M, Wang HG, Reed JC, et al. PAK1 phosphorylates the death agonist Bad and protects cells from apoptosis. Mol Cell Biol 2000;20:453–61.

[132] Jin S, Zhuo Y, Guo W, Field J. PAK1-dependent phosphorylation of RAF-1 regulates its mitochondrial localization, phosphorylation of BAD, and BCL-2 association. J Biol Chem 2005;280:24698–24705.

[133] Ye DZ, Jin S, Zhuo Y, Field J. Pak1 phosphorylates BAD directly at serine 111 in vitro and indirectly through Raf-1 at serine 112. PLoS One 2011;6:e27637.

[134] Jakobi R, Moertl E, Koeppel MA. p21-activated protein kinase g-PAK suppresses programmed cell death of BALB3T3 fibroblasts. J Biol Chem 2001;276:16624–16634.

[135] Tang Y, Zhou H, Chen A, Pittman R, Field J. The Akt proto-oncogene links Ras to Pak and cell survival signals. J Biol Chem 2000;275:9106–9.

[136] Vadlamudi R, Bagheri-Yarmand R, Yang Z, Balasenthil S, Nguyen D, Sahin A, et al. Dynein light chain 1, a PAK 1-interacting substrate, promotes cancerous phenotypes. Cancer Cell 2004;5:575–85.

[137] Mazumdar A, Kumar R. Estrogen regulation of Pak1 and FKHR pathways in breast cancer cells. FEBS Lett 2003;535:6–10.

[138] Tuazon P, Spanos W, Gump E, Monnig C, Traugh J. Determinants for substrate phosphorylation by p21-activated protein kinase (g-PAK). Biochemistry 1997;36:16059–16064.

[139] Fabian JR, Daar IO, Morrison DK. Critical tyrosine residues regulate the enzymatic and biological activity of Raf-1 kinase. Mol Cell Biol 1993;13:7170–9.

[140] Morin PJ. Beta-catenin signaling and cancer. Bioessays 1999;21:1021–30.

[141] Easwaran V, Song V, Polakis P, Byers. S. The ubiquitin–proteasome pathway and serine kinase activity modulate adenomatous polyposis coli protein-mediated regulation of beta-catenin-lymphocyte enhancer-binding factor signaling. J Biol Chem 1999;274:16641–16645.

[142] Li J, Mizukami Y, Zhang X, Jo WS, Chung DC. Oncogenic K-ras stimulates Wnt signaling in colon cancer through inhibition of GSK-3beta. Gastroenterology 2005;128:1907–18.

[143] Taurin S, Sandbo N, Qin Y, Browning D, Dulin NO. Phosphorylation of beta-catenin by cyclic AMP-dependent protein kinase. J Biol Chem 2006;281:9971–6.

[144] He H, Shulkes A, Baldwin GS. PAK1 interacts with beta-catenin and is required for the regulation of the beta-catenin signalling pathway by gastrins. Biochim Biophys Acta 2008;1783:1943–54.

[145] He H, Huynh. N, Liu. K, Malcontenti-Wilson C, et al. PAK1 knockdown inhibits β-catenin signalling and blocks colorectal cancer growth. Cancer Lett 2012;317:65–71.

[146] Hannigan GE, Leung-Hagesteijn C, Fitz-Gibbon L, Coppolino M, Radeva G, Filmus J, et al. Regulation of cell adhesion and anchorage-dependent growth by a new beta 1-integrin-linked protein kinase. Nature 1996;379:91–6.

[147] Yau CY, Wheeler J, Sutton K, Hedley D. Inhibition of integrin-linked kinase by a selective small molecule inhibitor, QLT0254, inhibits the PI-3K/PKB/mTOR, Stat3, and FKHR pathways and tumor growth, and enhances gemcitabine-induced apoptosis in human orthotopic primary pancreatic cancer xenografts. Cancer Res 2005;65:1497–504.

[148] Acconcia F, Barnes C, Singh R, Talukder A, Kumar R. Phosphorylation-dependent regulation of nuclear localization and functions of integrin-linked kinase. Proc Natl Acad Sci USA 2007;104:6782–7.

[149] Lee SL, Hsu EC, Chou CC, Chuang HC, Bai LY, Kulp S, et al. Identification and characterization of a novel ILK inhibitor. J Med Chem 2011;54:6364–74.

[150] Hibi M, Lin A, Smeal T, Minden A, Karin M. Identification of an oncoprotein- and UV-responsive protein kinase that binds and potentiates the c-Jun activation domain. Genes Dev 1993;7:2135–48.

[151] Bagrodia S, Dérijard B, Davis RJ, Cerione RA. Cdc42 and PAK-mediated signaling leads to Jun kinase and p38 mitogen-activated protein kinase activation. J Biol Chem 1995;270:27995–27998.

[152] Brown J, Stowers L, Baer M, Trejo J, Coughlin S, Chant J. Human Ste20 homologue hPAK1 links GTPases to the JNK MAP kinase pathway. Curr Biol 1996;6:598–605.

[153] Nheu T, He H, Hirokawa Y, Tamaki K, et al. The K252a derivatives, inhibitors for the PAK/MLK kinase family selectively block the growth of RAS transformants. Cancer J 2002;8:328–36.

[154] Maroney A, Glicksman M, Basma A, Walton K, et al. Motoneuron apoptosis is blocked by CEP-1347 (KT 7515), a novel inhibitor of the JNK signaling pathway. J Neurosci 1998;18:104–11.

[155] Dan C, Nath N, Liberto M, Minden A. PAK5, a new brain-specific kinase, promotes neurite outgrowth in N1E-115 cells. Mol Cell Biol 2002;22:567–77.

[156] Ha UH, Lim JH, Kim HJ, Wu W, Jin S, Xu H, et al. MKP1 regulates the induction of MUC5AC mucin by Streptococcus pneumoniae pneumolysin by inhibiting the PAK4–JNK signaling pathway. J Biol Chem 2008;283:30624–30631.

[157] Cotteret S, Jaffer Z, Beeser A, Chernoff J. Pak5 localizes to mitochondria and inhibits apoptosis by phosphorylating BAD. Mol Cell Biol 2003;23:5526–39.

[158] Huser M, Luckett J, Chiloeches A, Mercer K, Iwobi M, Giblett S, et al. MEK kinase activity is not necessary for Raf-1 function. EMBO J 2001;20:1940–51.

[159] Greer E, Dowlatshahi D, Banko M, Villen J, et al. An AMPK–FOXO pathway mediates longevity induced by a novel method of dietary restriction in C. elegans. Curr Biol 2007;17:1646–56.

[160] Morrison H, Sherman LS, Legg J, Banine F, Isacke C, Haipek C, et al. The NF2 tumor suppressor gene product, merlin, mediates contact inhibition of growth through interactions with CD44. Genes Dev 2001;15:968–80.

[161] Druker BJ. Translation of the Philadelphia chromosome into therapy for CML. Blood 2008;112:4808–17.

[162] Kim EJ, Zalupski MM. Systemic therapy for advanced gastrointestinal stromal tumors: beyond imatinib. J Surg Oncol 2011;104:901–6.

3 Natural or Synthetic Therapeutics That Block PAKs

Hiroshi Maruta[1], Shanta M. Messerli[2], and Ramesh K. Jha[3]

[1]NF/TSC Cure Organisation, Brunswick West, Melbourne, Australia, [2]Marine Biological Laboratory, Woods Hole, MA, USA, [3]Bioscience Division, Los Alamos National Laboratory, Los Alamos, NM, USA

Abbreviations

NF	neurofibromatosis
TSC	tuberous sclerosis
DN	dominant negative
AID	autoinhibitory domain
BBB	blood–brain barrier
CTCL	cutaneous T-cell lymphoma
PTCL	peripheral T-cell lymphoma
HDAC	histone deacetylase
CA	caffeic acid
CAPE	caffeic acid phenethyl ester
ARC	artepillin C
KO	knockout
CDK	cyclin-dependent kinase

3.1 Introduction

3.1.1 The Dawn of PAK1–3 Inhibitor Development

Before we discuss in detail several distinct cell-permeable compounds that could be potentially useful as therapeutics for PAK1-dependent diseases, for educational purposes we will introduce two "ancestral" PAK1 inhibitors that were first developed more than a decade ago as useful laboratory reagents for studying the potential role of PAK1–3 (group 1 of the PAK family) in cell culture.

3.1.2 The DN Mutant of PAK1 (cDNA)

In the late 1990s, the first PAK1-specific inhibitor was generated in order to study if PAK1 is essential for the growth of the so-called RAS cancers, which are caused

PAKs, RAC/CDC42 (p21)-activated Kinases. DOI: http://dx.doi.org/10.1016/B978-0-12-407198-8.00003-5

by the oncogenic mutations of the GTPase RAS. They include more than 90% of pancreatic cancers and 50% of colon cancers, and represent more than 30% of all human cancers [1]. Since the related GTPases RAC and CDC42 are activated by the oncogenic mutants of RAS through a lipid kinase called PI-3 kinase, and PAK1 is activated by RAC/CDC42 directly *in vitro*, Jeff Field's group in Philadelphia asked if RAS cancer cells (transformed fibroblasts) require PAK1 for their malignant (anchorage-independent) growth. Their so-called PAK1-specific inhibitor is a kinase-negative dead mutant called dominant negative (DN) in which the kinase domain is mutated. It is no longer active as a kinase but is still able to bind either its substrates, such as the kinases RAF and LIM kinase, or its activators, such as RAC/CDC42, PIX, and the Tyr kinase ETK. In other words, it could block the essential interaction of PAK1 with either its substrates or its activators. This DN mutant could be overexpressed in RAS-transformed cells through a cDNA encoding this mutant. In 1997, they found that a RAS-transformed fibroblast overexpressing this DN mutant is no longer able to grow in soft agar, meaning that PAK1 is essential for the anchorage-independent growth of this RAS transformant, although PAK1 is not required for the anchorage-dependent growth of normal fibroblasts [2]. Furthermore, this DN mutant overexpressing RAS transformants failed to grow *in vivo* in nude mice. This finding suggests that RAS cancers, representing more than 30% of all human cancers, require the kinase PAK1, and cell-permeable anti-PAK1 drugs could eradicate all these PAK1-dependent cancers without any adverse side effect, if such compounds were to become available in the future. Unfortunately, however, this DN-mutant cDNA is not cell-permeable *per se*, and in general gene therapy of cancers using tumor suppressor genes such as DN mutants is not yet practical clinically, mainly because unlike reovirus infection, which selectively kills all RAS transformants, the efficiency (and selectivity) of current DNA transfection technology is not high enough to target all RAS cancer cells in each patient. The potential problem associated with the reoviral therapy is that repeated viral infection would cause the development of antibodies against this virus in RAS cancer patients so that the therapeutic effect of reovirus would be rather short lived.

3.1.3 TAT/WR-PAK18 (Peptide)

Our first opportunity to generate a cell-permeable anti-PAK1 drug was prompted by the 1998 findings of Ed Manser's group in Singapore. They discovered a new SH3 adaptor protein called PIX that binds the Pro-rich domain (residues 186–203) of PAK1 [3]. This PIX–PAK1 interaction is required for the activation of PAK1 in cells, because microinjection of 18 amino acid PAK1 peptide (PAK18) corresponding to this PIX-binding motif blocks the PIX–PAK1 interaction and the RAS/RAC-induced membrane ruffling of fibroblasts [4]. Since this membrane ruffling (actomyosin-based membrane dynamics at the leading edge of cells) is closely linked to RAS transformation and loss of gap junction, which requires PAK1, we thought that the peptide PAK18 could block RAS transformation if it is converted to a cell-permeable derivative. In the past, we were successful in generating another cell-permeable peptide that blocks RAS transformation. This peptide is called WR-ACK42, and it links the cell-permeable peptide vector WR of 16 amino acids to ACK42, the CDC42 binding site (residues 504–545) of the Tyr kinase ACK [5]. This WR-ACK42 is a DN mutant of ACK and selectively blocks the CDC42–ACK interaction essential for the activation of ACK. Since WR-ACK42 completely blocks the RAS transformation, it became clear that RAS cancer cells require not only the kinase PAK1 (and its activator PIX), but also the kinase ACK (and its activator CDC42) for their growth [5]. By a similar approach, we generated the cell-permeable peptide WR-PAK18 [6]. WR-PAK18 (10 μM) completely blocks not only RAS/RAC-induced membrane ruffling but also RAS-induced malignant (anchorage-independent) growth, without affecting the normal (anchorage-dependent) growth of fibroblasts at all [6].

However, WR-PAK18 has at least three potential problems for its clinical use. The first is that the vector WR is a highly basic (Arg-rich) peptide and could be susceptible to a proteolytic cleavage in the circulation, although TAT-PAK18 (in which WR is replaced by another highly basic peptide vector TAT) appears to be effective even *in vivo* (animal experiments) [7]. The second is that peptide synthesis in general is rather expensive. A third problem could be that since both WR and TAT are "foreign" peptides, repeated treatment of patients with these potential immunogens might cause an antibody production against these vectors in the circulation. Thus, if available, small (cell-permeable) chemical compounds that selectively inactivate PAK1 would be far better and more practical than these peptides for clinical application.

A few years ago, a third PAK1–3 inhibitor called IPA-3 was developed by Jeff Peterson's group at Fox Chase Cancer Center (FCCC) [8] and has been widely used mainly as a laboratory reagent for cell culture studies. This is a synthetic dinaphthyl-disulfide compound that binds the autoinhibitory domain (AID) of PAK1–3, blocking the activation of PAK1–3 by GTPase (RAC/CDC42). Unfortunately, however, its IC_{50} is rather high (around 30 μM) in cells, and this disulfide-bonded dimer is inactivated by glutathione or other reducing agents (antioxidants) *in vivo*. Thus, it is very unlikely that IPA-3 is useful for clinical application, unless this disulfide bond is replaced by a non-reducible C-C (or C=C) bond if possible, which would boost its anti-PAK1–3 activity by around 1000 times, as well as improving its cell permeability through a further substantial chemical modification including the aminohexyl side chain, perhaps replacing hydroxyl groups, for instance, to give a more positive charge so that it can more efficiently cross the negatively charged cell membranes.

3.2 Synthetic Chemical Therapeutics

In an attempt to explore effective therapeutics for PAK1-dependent diseases, we have taken two distinct approaches to develop or identify anti-PAK1 compounds.

The first approach is to identify the direct PAK1-specific inhibitor among the known kinase inhibitors. The second approach is to identify any chemical compounds that somehow directly or indirectly block the activation of PAK1 in cells.

3.2.1 CEP-1347 and ST3009/ST3010

The antibiotic K-252a is known to be among nonspecific kinase inhibitors, but its specific derivative, called CEP-1347 (Figure 3.1), developed jointly by Kyowa Hakko and Cephalon, turned out to inactivate the kinase JNK somehow, which is essential for the apoptosis of neuronal cells [9]. Thus, it was thought to be potentially useful for the therapy of Parkinson's disease (PD). However, JNK is not the direct target of this drug. Instead, we found that both PAK1 and MLKs (mixed-lineage kinases), which are essential for the activation of JNK, are the direct and specific targets of this drug [10]. Interestingly, both PAK1–3 and MLK family kinases are RAC/CDC42-dependent kinases and uniquely form an inactive homodimer [11] as has been discussed in detail in Chapter 1. Thus, their ATP-binding pocket is much larger than any of other kinase families such as PKC and PKA. K-252a is a natural ATP antagonist and nonselectively inhibits most kinases. However, CEP-1347, generated by linking a bulky hydrophobic side chain at positions 3 and 9 of K-252a, inhibits selectively, only affecting PAK1/PAK3 and MLKs, which have an extra, larger ATP pocket [10]. The IC_{50} of CEP-1347 for MLKs and PAK1 are around 20 nM and 1 μM, respectively. When RAS-transformed (or breast cancer) cells are treated with CEP-1347 at 20 nM and 1 μM, only at 1 μM is their growth strongly inhibited, clearly indicating that these cancer cells require PAK1, but not MLKs, for their malignant growth [12]. The only potential problem with this drug is that its IC_{50} is too high for clinical application. An IC_{50} around 1 nM would be more desirable in a practical PAK1 inhibitor that is useful for cancer therapy.

Thus, we began hunting such a potent PAK1 inhibitor. Around 2001, in collaboration with a German group, we found one called ST2001 [10]. It is a natural staurosporine (ST) derivative, 3-OH ST, produced by a specific marine organism [13]. It inhibits both PAK1 and PKC with the IC_{50} around 1 nM, while the parental ST inhibits PAK1 and PKC with the IC_{50} around 50 and 10 nM, respectively [10]. According to the 3D structure of ST2001/PAK1 complex (PDB ID 2HY8), the tight hydrogen bond between 3-OH and the backbone of PAK1 at Leu 347 in the ATP-binding pocket appears to increase their affinity by 50-fold. In theory, if position 9 is modified by positively charged bulky side chains such as hexylamine, the anti-PKC activity is selectively abolished without affecting the anti-PAK1 activity, and

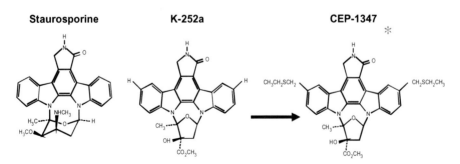

Figure 3.1 CEP-1347 derived from K252a.

the resulting, ST3009 (Figure 3.2), would be water soluble and more efficiently penetrate negatively charged cell membranes. Furthermore, the 3D image of ST2001/PAK1 complex (Figure 3.3) suggests that a huge cavity (indicated by an orange arrow) is available beyond the position 9. Thus, if this amino-alkyl side chain at position 9 or 10 could be further rotated clockwise by 60°, it would nicely fit in this spacious straight tunnel without any restriction.

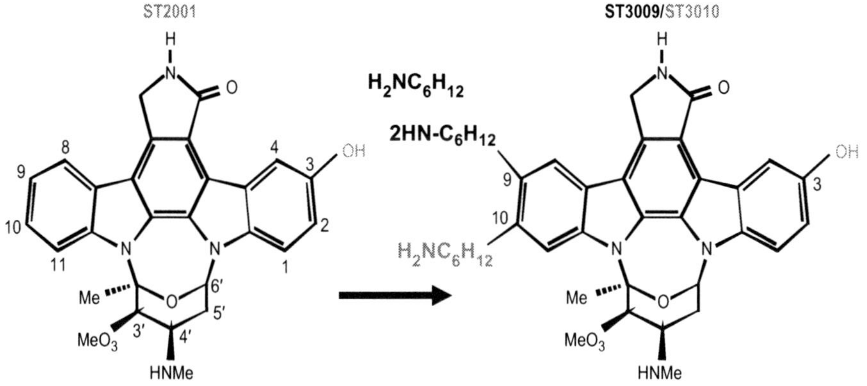

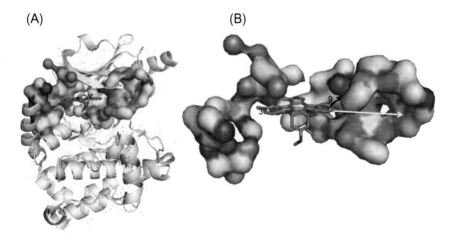

Figure 3.2 Conversion of ST2001 to ST3009.

Figure 3.3 ST2001 in ATP-binding pocket of PAK1. (A) ATP-binding pocket of PAK1 from PDB 3Q53 is shown in green. ST2001 is magenta and PAK1 sites bound to ST2001 are cyan. (B) Closer view of the ATP-binding pocket with ST2001. Any bulky substitution at position 9 will run into a dead end and hence is shown with a red arrow, meaning "stop." However, any substitution that corresponds to 60° clockwise has a spacious "tunnel" to accommodate molecules such as aminohexyl side chains, hence these are shown with an orange arrow, meaning "ready to go." (For interpretation of the references to color in this figure legend, the reader is referred to the web version of this book.)

Unfortunately, however, this unique marine organism suddenly disappeared from the coastline of Guam (perhaps due to recent global warming) as soon as we launched this ST3009 project. Although ST2001 can be synthesized by a chemical reaction from ST, the majority of the reaction products would be 3, 9-OH ST, which is no longer active to inhibit PAK1, and the yield of 3-OH ST would be minute. Thus, the chemical synthesis of ST2001 is not economically feasible. The remaining option would be an enzymatic hydroxylation of ST at position 3 to produce ST2001. During the indolocarbazol ring formation of ST biosynthesis, two Trp molecules form a dimer, and Trp hydroxylase could attack either position 3 or 9 to produce 3-OH or 9-OH ST [14,15]. Thus, if the enzymatically synthesized ST2001 becomes available some days in the future, it would be possible to make our dream compound, ST3009 or ST3010, a "magic bullet" for cancer therapy. Alternatively, if we could manage to attach the aminohexyl side chain at position 10 of ST somehow, perhaps we could chemically hydroxylate at position 3 selectively.

3.2.2 Combination of Two Tyr Kinase Inhibitors

Around the turn of this century, we found two distinct (independent) Tyr kinase pathways that lead to full activation of PAK1 in cells. One pathway involves an SRC family Tyr kinase, and the other involves another Tyr kinase called ETK that directly phosphorylates PAK1 [7,16,17]. The treatment of RAS-transformed cells with the SRC kinase inhibitor PP1 at 10 nM inhibits both PAK1 activity and the cells' growth by 50%, indicating that one of the SRC family kinases is essential for both PAK1 activation and RAS transformation [16]. Interestingly, the combination of PP1 and AG1478, an ErbB1- (EGF receptor-) specific inhibitor, does not enhance PP1-induced inactivation of PAK1, although AG1478 alone inactivates both PAK1 and cell growth only by 50%, clearly indicating that this SRC family kinase acts downstream of ErbB1 to activate PAK1 [16]. Moreover, a Chinese group at Fudan University in Shanghai recently found that PP2, a PP1-related inhibitor of SRC family kinases, inhibits the phosphorylation and inactivation of a tumor suppressor called HT1 (trihydrophobin 1) that normally inhibits PAK1 directly [17]. These findings together strongly suggest, if not proven as yet, that PP1/PP2-induced inactivation of PAK1 involves SRC-induced HT1 phosphorylation at Tyr 6, and the reactivated HT1 in turn blocks PAK1 and the growth of RAS-transformed cells (Figure 3.4).

The treatment of RAS-transformed cells with AG 879, an ETK-specific inhibitor, at around 5 nM inhibits both PAK1 activity and cell growth by 50%, indicating that ETK is also essential for both PAK1 activation and RAS transformation [18]. When the same cancer cells are treated with both PP1 and AG 879, the PAK1 is completely inactivated and RAS transformation is blocked by 100% [18]. Thus, two distinct Tyr kinases, an SRC family kinase and ETK, are involved independently in both PAK1 activation and RAS transformation (see Figure 3.4). Since AG 879 is water insoluble, we created a water-soluble (hexylamine) derivative called GL-2003 and found that the combination of PP1 and GL-2003 (each 20 mg/kg, i.p., twice a week) almost

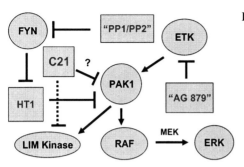

Figure 3.4 SRC-HT1-PAK1 signaling.

completely suppresses the growth of human pancreatic and breast cancer xenografts in mice without any adverse effect [19].

Although the responsible direct target of PP1 among SRC family kinases remained to be clarified until very recently, FYN appears to be among its likely targets because the IC_{50} of PP1 for blocking PAK1 is around 10 nM, which is very close to the IC_{50} for FYN [20]. In 2011, Mitchell Denning's team at Loyola University in Chicago found that the oncogenic RAS upregulates FYN mRNA dramatically (>100-fold) and its kinase activity through AKT, and that either siRNA for FYN or the inhibitor PP2 strongly inhibits the PAK1-dependent metastasis/invasion of human breast cancers [21]. In other words, two distinct Tyr kinases, FYN and ETK, are required for the full activation of PAK1 in RAS-transformed cells (see Figure 3.4).

3.2.3 PF3758309 (UnPAK309 or PanPAK309)

In 2010, Brion Murray's team at Pfizer Oncology in San Diego developed a new Pan-PAK inhibitor called PF3758309 (UnPAK309 or PanPAK309, for those who cannot remember such monotonous seven digits!) that selectively inhibits PAK family kinases including PAK1 and PAK4 with the IC_{50} between 5 and 15 nM in cells [22]. *In vivo*, using cancer xenografts in nude mice, this compound suppresses the growth of several PAK1/PAK4-dependent human cancers such as breast, colon, and lung cancers as well as melanoma at daily doses around 20 mg/kg. Since its derivatives pass the blood–brain barrier (BBB), it is likely that UnPAK309 could be useful for the treatment of brain tumors such as glioma and neurofibromatosis/tuberous sclerosis (NF/TSC) tumors in the future.

However, as UnPAK309 clinical trials have just started, it will take several years to clarify whether this potent PAK inhibitor could become available on the market, because this drug strongly inhibits PAK4, which is essential for cardiovascular development (Figure 3.5), and like AKT-null mice, PAK4-null mice are embryonic lethal, mainly due to severe heart failure [23]. Furthermore, unlike most of the anti-PAK1 drugs that activate the tumor-suppressive kinase AMPK, UnPAK309 directly inhibits AMPK [22], which is essential for the activation of glucose transporter GLUT-4, required for the cellular uptake of glucose from the bloodstream, and therefore it might cause type 2 diabetes as well. (This is discussed in more detail later.)

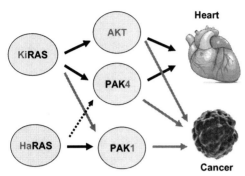

Figure 3.5 PAK4 versus PAK1. Like KiRAS-null and ATK-null mice, PAK4-null mice are embryonic lethal, mainly due to a severe heart failure. In addition, KiRAS, ATK, and PAK4 are responsible for cancer development. Like HaRAS-null mice, PAK1-null mice are perfectly healthy, but hyperactivation of HaRAS or PAK1 is responsible for cancer development.

3.3 Natural PAK1 Blockers

3.3.1 FK228

Around 1994, a potent antibiotic ring peptide called FR901228 (now called FK228 for short) was developed by Fujisawa Pharmaceuticals (precursor to Astellas Pharma) in Japan. FK228 inhibits the growth of RAS-transformed cells with the IC_{50} around 1 nM [24]. It was isolated from culture of the soil bacteria *Chromobacterium viola-ceum*. Later its direct target was found to be histone deacetylase (HDAC). FK228 activates a specific set of genes including p21/WAF1, an inhibitor of cyclin-dependent kinase (CDK), by histone acetylation of chromatin [25]. We found that FK228 eventually inactivates PAK1 and blocks the growth of breast cancer cells with the IC_{50} far below 1 nM [12]. Since another oncogenic kinase, AKT, is also blocked by FK228 [26], it is most likely that this drug somehow inactivates PI-3 kinase, which activates both AKT and PAK1. However, a tumor suppressor called PTEN, which antagonizes PI-3 kinase by dephosphorylating phosphatidylinositol 3-phosphate (PIP3), is not activated by FK228. Instead, FK228 upregulates Rap1, a RAS antagonist [27], which blocks the RAS-induced activation of both PI-3 kinase and RAF1.

FK228 is currently in clinical trials (phases 2–3), mainly for rare malignant lymphomas called cutaneous T-cell lymphoma (CTCL) and peripheral T-cell lymphoma (PTCL) [28], and in the summer of 2011, Celgene successfully marketed this inject-able drug under the brand name Istodax. Although this drug is so far the most potent anticancer drug (and PAK1 blocker), it is unable to pass the BBB. Thus, FK228 would be ineffective on brain tumors such as glioma and NF/TSC-associated tumors. We shall discuss these brain tumors in detail in Chapter 5.

Nevertheless, using FK228 and a few other PAK1 blockers, we and others found that more than 70% of all human cancers, in particular solid tumors, require PAK1 for their growth. These PAK1-dependent cancers include pancreatic and colon can-cers (the so-called RAS cancers), breast and prostate cancers (representing around 30% of female and male cancers, respectively), ovarian, cervical, thyroid, stomach, lung, and liver cancers as well as glioma, melanoma, multiple myeloma (MM), NF1-deficient malignant peripheral nerve sheath tumor (MPNST), and NF2-deficient mesothelioma. For details, see Chapter 2. Among rare tumors, CTCL, NF and TSC tumors, and Cowden syndrome (PTEN-deficient) tumors are also PAK1-dependent tumors. *NF*1 and *NF*2 gene products are tumor suppressors, a RAS GAP (attenuator), and a PAK1 inhibitor called merlin, respectively, and their dysfunction causes NF type 1 and type 2 respectively [29,30]. Overexpression of either the *NF*1 or *NF*2 gene reverses RAS transformation [31,32]. For detail, see Chapter 5 on PAK1-dependent brain tumors and other neuronal disorders.

One of the major reasons why these solid tumors require PAK1 is that these tumors require angiogenesis (new blood vessel formation around tumors) for their growth, and PAK1 is essential for this tumor-induced angiogenesis [33]. However, PAK1 is not essential for normal angiogenesis during embryogenesis because PAK1-deficient mice are perfectly normal [34]. Thus, unlike thalidomide, natural PAK1 blockers in general do not cause any teratogenesis. In addition, as described in Chapter 2 in detail, PAK1 activates LIM kinase, which is essential for the metastasis/invasion of malignant tumors [35]. Thus, unlike conventional anticancer drugs (DNA/RNA/microtubule poisons), a single PAK1 blocker would suppress many aspects of malignancy of these solid tumors. In addition, most of these PAK1 blockers would not cause any of the side effects (hair loss, immunosuppression, or loss of appetite) often caused by conventional anticancer drugs, such as cisplatin, 5-FU, taxol, and gemcitabine, which block the growth of fast-growing normal cells such as bone marrow and hair cells.

3.3.2 Propolis

Although FK228 and UnPAK309 are potent PAK1 blockers, it would take 7–10 more years for them to become available on the market for major solid cancers. Thus, at present none of them is useful for terminal cancer patients such as those who are currently suffering from pancreatic cancers and whose life expectancy is only 3–4 months. To overcome such a grim reality, around 2005, we started exploring PAK1 blockers extensively among natural products that are freely available on the market. So far an old bee product called propolis has turned out to be the most potent and reliable natural product that blocks PAK1.

Why did we get interested in a bee product in the first place? There is an old German saying: beekeepers never get cancer. More precisely, only 1 in 3000 beekeepers suffer from cancer, while 1 in 3 nonbeekeeping people get cancer during their lifetimes. In other words, beekeepers are 1000 times more resistant to cancer than are ordinary people. Why? Honey, royal jelly, and propolis are three major bee products, but the first two have no significant anticancer properties. Has propolis any anticancer properties? That is what Dezider Grunberger (1922–1999) at Columbia University asked himself around the mid-1980s. Then, in collaboration with Koji Nakanishi, an eminent organic chemist at the same university, his team started testing anticancer properties of an extract of propolis from Israel and found that a compound called caffeic acid phenethyl ester (CAPE) (Figure 3.6) in this ethanol extract kills cancer cells, without affecting the growth of normal cells [36]. In other words, it is most likely that propolis makes beekeepers so resistant to cancer.

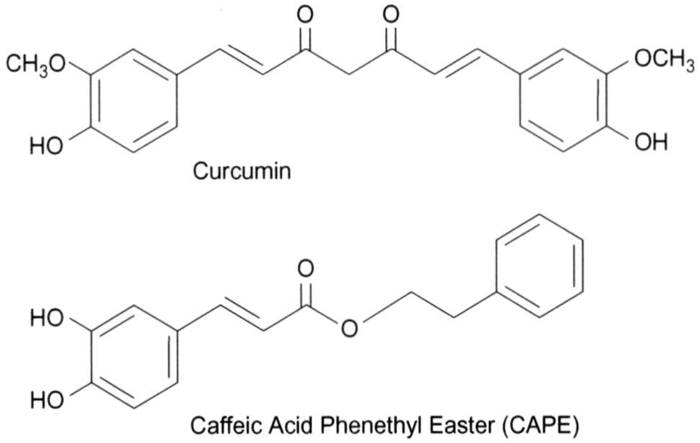

Figure 3.6 CAPE and curcumin.

Propolis is a 100 million-year-old bee-made antibiotic that protects bee larvae living in honeycombs from harmful parasites such as bacteria, fungi, and viruses. Since the ancient Egyptian era, more than 4000 years ago, people have used propolis for treating wounds and inflammation. In addition, they used propolis to prepare the bodies of deceased royalty for mummification. Hippocrates (460–370 BC), the father of medicine in the ancient Greece, was the first person who called this healing bee product that guards the entrance to the honeycomb *propolis*, combining the Greek words for "in front of" (*pro*) and "city" (*polis*).

Each honeycomb is made of wax (fat) and propolis, and the major sources of propolis are young buds of poplar trees, willow, and a few other specific trees depending on the climate. There are at least three distinct types of propolis, distinguished by their major anticancer ingredients. Propolis prepared in temperate zones such as Europe, the Far East, North America, and Oceania is based on CAPE. However, propolis in Brazil contains no CAPE.

Instead, Brazilian green propolis contains artepillin C (ARC) as the sole anticancer ingredient [37]. Brazilian red propolis contains triterpenes, instead of ARC, as the major anticancer ingredients [38]. Interestingly, despite a huge difference in the major anticancer ingredients, all three types of propolis stop the PAK1-dependent growth of pancreatic cancer cells without affecting normal cell growth [36–38]. These observations suggest that there must be a common mechanism underlying their anticancer properties at the molecular level.

The first clue emerged around 2005. CAPE is derived from caffeic acid (CA) by conjugating with phenethyl alcohol through an enzymatic or chemical reaction (Figure 3.7). A Japanese group found that CA downregulates the GTPase RAC, which is the direct activator of PAK1 [39]. In other words, CA blocks PAK1, and it is most likely that CAPE does the same. We found that ARC indeed blocks PAK1 selectively, without affecting another oncogenic kinase, AKT [40]. Furthermore, both ARC and CAPE activate Hsp16 gene in *Caenorhabditis elegans*, which is inactivated by PAK1 [41], clearly confirming that CAPE also blocks PAK1. As described

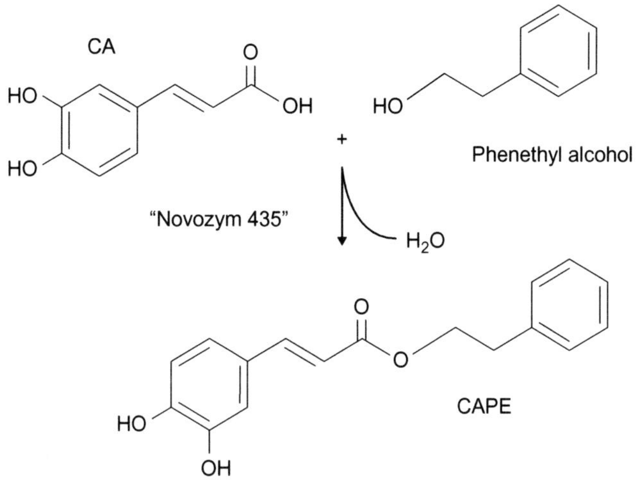

Figure 3.7 Enzymatic synthesis of CAPE.

in Chapter 5 in detail, both CAPE- and ARC-based propolis extracts completely suppress the PAK1-dependent growth of *NF*2-deficient schwannoma xenografts in mice without any adverse effect [40,42].

3.3.3 Antimalarial Drugs

As one of us describes in Chapter 4 in detail, like HIV and influenza viruses, malaria parasites require PAK1 in host cells such as erythrocytes for their infection [43]. In other words, it is quite possible that some currently used antimalarial drugs block PAK1 to cure malaria. At least several of them have indeed been found to be anti-PAK1 drugs: artemisinin, chloroquine, quinidine, curcumin, resveratrol, glaucarubinone, berberine, propolis and FK228. For a recent example, among the most widely used antimalarial drugs that are effective against multidrug-resistant malaria, artemisinin derivatives such as artemether and artesunate indeed block PAK1 [44], and completely suppress the PAK1-dependent growth of human pancreatic cancer xenografts in mice [45].

Back in 1947, it was shown that an extremely bitter extract from the bark of a tree family called *Simaroubaceae* administered at 1 mg/kg is sufficient to suppress malaria in chickens. This extract has been used by native people in Amazon jungles to treat malaria for many centuries. In 1981, it was found that the major antimalarial ingredient in this extract is a triterpene/quassinoid called glaucarubinone ([46], Figure 3.8). Interestingly, in 1977 this compound was demonstrated to inhibit the growth of murine leukemia cells with the IC_{50} around 1 μM [47]. Furthermore, in 1998, a group led by Fred Valeriote and Paul Grieco demonstrated its therapeutic effect on a solid mammary tumor developed in mice [48]. However, the molecular mechanism underlying either its antimalarial or anticancer action remained unclear until recently. In 2009, John Beutler's group at the National Cancer Institute (NCI)

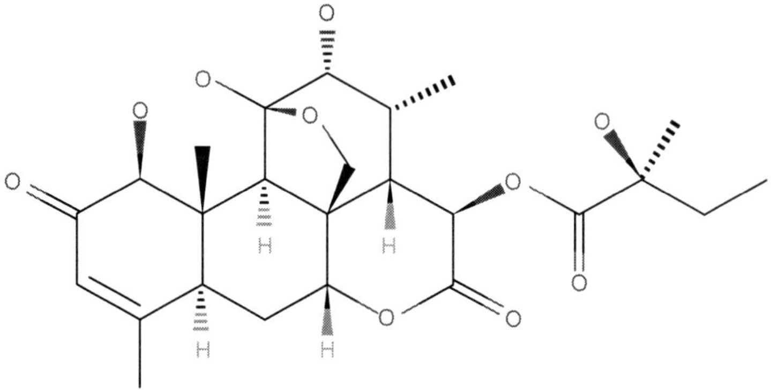

Figure 3.8 Glaucarubinone.

found that glaucarubinone blocks an oncogenic transcription factor called AP-1 with the IC_{50} around 20 nM [49]. Since either PAK1 or AKT is essential for the activation of AP-1 [50,51], it is likely that this quassinoid is a PAK1/AKT blocker. Furthermore, in 2011 a German group at Jena University reported that this compound significantly extends the lifespan of *C. elegans*[52], supporting the notion that it blocks PAK1 or AKT or their common upstream activator PI-3 kinase, improving the health conditions in this worm. In 2010, a Swiss group at Zurich University Hospital reported that a potent semisynthesized quassinoid derivative called NBT-272 (a derivative of peninsularinone) blocks both AKT and PAK/ERK, and inhibits the growth of human medulloblastoma cells with the IC_{50} ranging from 5 to 15 nM [53]. However, it turned out that PI-3 kinase is not its direct target, indicating that NBT-272 targets further upstream signal transducer(s) such as RAS and ErbB1 (EGF receptor). This derivative was originally created by Paul Grieco's group in 1998, but later renamed NBT-272 and further developed by NaPro Biotherapeutics/Tapestry Pharma. The maximum tolerable dose is around 1 mg/kg for mice.

Finally, in 2012, we confirmed directly that glaucarubinone blocks PAK1 in human colon cancer cells and inhibits the PAK1-dependent growth of both *NF2*-deficient Schwannoma cells with the IC_{50} around 60 nM (Figure 3.9) and colon cancer cells with the IC_{50} ranging between 65 and 105 nM (H He, J Beutler, G Baldwin, unpublished observation). Thus, it is most likely that this natural quassinoid also inactivates both PAK1 and AKT, probably by targeting PI-3 kinase or its further upstream signal transducer. We are planning to test the therapeutic effect of glaucarubinone or NBT-272 on human colon cancer/schwannoma xenografts in mice in the near future.

Chloroquine has been widely used for the treatment of both malaria and rheumatism. In 2011, Alec Kimmelman's group at Dana Farber Cancer Center found that this drug (60 mg/kg, i.p., daily) strongly suppresses the autophagy and growth of human pancreatic cancer xenografts in mice [54] and recently started clinical trials for terminal pancreatic cancers using a derivative called hydroxychloroquine (HC) that causes fewer side effect(s) than chloroquine during long-term repeated

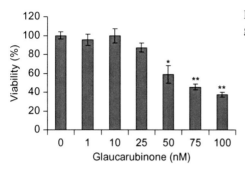

Figure 3.9 Glaucarubinone inhibits the growth of schwannoma cells.

treatment. Since PAK1 is essential not only for treatment of malaria and inflammatory diseases such as rheumatism, but also for the growth of RAS transformants such as pancreatic and colon cancers [2,7,18]; it is almost certain that chloroquine and HC also block PAK1 somehow.

3.3.4 Chinese (Sichuan) Peppercorn Extract

Hua Jiao (Zanthoxyli Fructus), a peppercorn from Sichuan Province in China, has been long used as a special seasoning in preparing a Sichuan dish called *mapo tofu*, a spicy bean curd soup. Around 2005, we found that a 70% ethanol extract of this red peppercorn selectively blocks PAK1, without affecting another oncogenic kinase, AKT, and inhibits the growth of RAS transformants and NF1-deficient MPNST tumor cells with the IC_{50} around 10 μg/ml [55]. This extract has no effect on normal cell growth. *In vivo* (using xenografts in mice), this extract (100 mg/kg, i.p., twice a week) strongly suppresses the growth of *NF*1-deficient breast cancer (MDA-MB 231 cell line), in which both RAS and PAK1 are abnormally activated, without any adverse effect on mice [55]. Currently John Beutler's group at NCI is isolating the major anti-PAK1 ingredient (called pepperin) in this extract and determining its chemical structure.

3.3.5 Curcumin

Among the natural antimalarial substances, the CAPE in propolis and curcumin, found in turmeric powder (the spicy yellow ingredient in Indian curry), are structurally very similar (see Figure 3.6), and as expected, both polyphenols were found to block PAK1 and activate AMPK [41,56–58]. Like CAPE, curcumin suppresses the growth of many PAK1-dependent cancer cells *in vitro*, but neither CAPE nor curcumin alone has ever been used clinically, mainly because of its poor bioavailability (water insolubility). However, around 2005, Razella Kurzrock's group at M.D. Anderson Cancer Center solubilized curcumin by encapsulating curcumin in liposomes and successfully started demonstrating its high efficacy in suppressing the PAK1-dependent growth of human colon and pancreatic cancer xenografts in mice: curcumin in liposomes (20–40 mg/kg) suppresses the growth of these cancers by 50% [59,60]. Furthermore, in 2009, an Indian group led by Sarasija Suresh at Nootan Dental College in Bangalore found that a chemically synthesized β-dextrin derivative called hydroxypropyl-β cyclodextrin (CD) was the best CD for solubilizing curcumin effectively and showed both its antiangiogenic and anti-inflammatory effects *in vivo*[61]. Thus, in the future (probably in a decade or so), curcumin in either liposomes or CDs may become available on the market for the therapy of these cancers and NF tumors as well.

3.3.6 Emodin

This anthraquinone derivative is the major anticancer/anti-inflammatory ingredient in roots of Da Huang, a Chinese traditional medicine, also called Turkish rhubarb. Its anticancer property was established in the late 1990s *in vivo* by several groups including Huan MienChie's group at M.D. Anderson Cancer Center in Houston. Emodin alone (40 mg/kg, i.p., twice a week) significantly reduces the size of human breast cancer grafted in mice, and its combination with taxol causes a synergistic effect [62]. Emodin turned out to suppress several PAK1-dependent cancers such as breast cancer, melanoma, and glioma. However, the major target of emodin was originally thought to be the Tyr kinase ErbB2.

In 2005, however, Choon Nam Ong's group at Singapore National University found that Emodin inactivates PAK1 directly by blocking its interaction with RAC and CDC42 [61]. More interestingly, Emodin reduces the blood sugar level in type 2 diabetes mouse model by stimulating glucose uptake into cells [63], strongly suggesting, if not proving as yet, that like curcumin it eventually activates AMPK by inactivating either ErbB2 or PAK1, which in turn activates the glucose transporter GLUT-4.

3.3.7 Berberin

A traditional Chinese and American Indian herb called goldenseal (or yellow root) contains an anticancer and anti-inflammatory, yellow-colored, bitter-tasting alkaloid called berberine. Unlike the majority of anticancer products, berberine is relatively water soluble, and its tannic acid salt (berberine tannate) is tasteless and used clinically worldwide as an antibiotic for the therapy of infectious diseases such as malaria. In 1990, an NCI group found that berberine (0.1 mg/ml) induces a morphological differentiation of RAS-transformed teratocarcinoma cells *in vitro* [64]. However, the molecular mechanism underlying its anticancer action remained unknown until recently. In 2006, a French group at GlaxoSmithKline (GSK) found that berberine (100 mg/kg, daily) blocks both the synthesis and the accumulation of fats such as cholesterol in mice by activating the tumor-suppressive kinase AMPK [65]. In 2009, a Chinese group at Hong Kong University found that berberine inactivates both RAC and CDC42, thereby inactivating PAK1 and other PAKs [66]. Thus, like propolis and other anti-PAK1 products, this alkaloid would be useful for the treatment of PAK1-dependent cancers, NF and inflammatory diseases such as arthritis and asthma, as well as type 2 diabetes and obesity. However, its minimum effective daily dose for cancer/NF patients still remains to be determined.

Recently, a group led by Doug Kinghorn at Ohio State University synthesized a far more potent (8,8-dialkyl substituted) analog(s) from inexpensive starting materials such as berberine hemisulfate, which kills parasites such as trypanosoma and malaria with the IC_{50} between 10 and 40 nM *in vitro* but causes no side effect on normal host cells. This new berberine analog reduced leishmaniasis (liver parasitemia) in a mouse model by 50% even with a daily dose around 1 mg/kg, i.p. [67]. Although this analog was originally developed for the treatment of these infectious diseases, we hope it will also be useful for the treatment of other PAK1-dependent diseases such as solid tumors in the future.

3.3.8 Ivermectin

In an attempt to explore another anti-PAK1 drug among inexpensive and safe old drugs available on the market, we recently got interested in an old antiparasitic drug called ivermectin. After we completed the Bio 30 (CAPE-based propolis extract from New Zealand) project at Hamburg University Hospital (UKE) for a year [42], one of us (Dr. Maruka) spent a few months during the summer of 2007 at the University of Maryland in Baltimore (UMB) learning how to handle a tiny nematode (1 mm long) called *C. elegans*. The main purpose of this new project of ours was to develop an inexpensive and quick *in vivo* anti-PAK1 (anticancer) drug screening system based on this tiny and transparent worm, which lives only for 2–3 weeks. (Its generation time is only 2–3 days!). Unlike cloned immune-deficient mice such as nu/nu or SCID mice used for cancer xenografts, which cost $50 per head, a thousand nematodes cost almost nothing; they feed on bacteria such as *Escherichia coli* in a Petri dish. This worm consists of less than 1000 cells, including around 300 neuronal cells, 300 muscle cells, and 300–400 reproductive and digestive organ cells.

During that summer, we established a novel anti-PAK1 drug screening system using a special strain of this worm (CL2070) that carries a reporter gene called Hsp16-GFP [41]. In brief, we found that PAK1 normally suppresses the Hsp16 gene, which produces a heat-shock protein of 16 kDa after a brief heat-shock treatment. Thus, when the PAK1 gene is knocked out (KO), or the strain CL2070 is treated overnight with anti-PAK1 drugs such as CAPE or ARC, GFP, which is under the control of the Hsp16 gene promoter, is highly expressed, and the whole worm glows green under a blue light.

Interestingly, during this worm project, we found the very first phenotype of the strain RB689 in which PAK1 gene is KO; it has a very small litter size (number of eggs laid) compared with the wild-type worm. Either CAPE or ARC treatment of the wild-type worm causes a similar phenotype: a dramatic reduction of litter size. After finishing this nematode project, we visited La Trobe University in Melbourne, to see our old friend Warrick Grant, who is currently working with this nematode. Just in front of his office, we saw an old poster of his group from a few years ago, when he used to work in New Zealand for an agricultural research company. The abstract said that a sublethal dose of ivermectin reduces the litter size of this nematode by around 90%! It sounds like the anti-PAK1 drug effect. Ivermectin is an old antiparasitic antibiotic that mainly kills intestinal nematodes in humans and domestic animals. It was developed jointly by Merck and Kitasato Institute during the 1980s, and it selectively blocks the GABA receptor of nematodes but not the mammalian counterpart [68].

To test the potential anti-PAK1 activity of ivermectin in mammalian cells, we contacted a young Japanese oncologist, Tamotsu Sudo, who works at the Hyogo Cancer Center. In collaboration with his group, we found that ivermectin indeed inactivates PAK1, as the phosphorylation of the kinase RAF1 at Ser 338 by PAK1 is dramatically reduced in the presence of ivermectin with the IC_{50} around 5 μM, and the PAK1-dependent growth of both human ovarian cancer and *NF2*-deficient Schwannoma cells is strongly inhibited by this old drug [69].

More interestingly, around 2004, a Russian group in Moscow reported that ivermectin completely suppresses the growth of a few human cancers in mice, such as melanoma xenografts, with the daily dose of around 3 mg/kg [70], which is 15 times higher than the daily dose for killing intestinal nematodes (0.2 mg/kg). However, its anticancer mechanism at molecular levels remained to be clarified until recently when it was eventually unveiled by us. Since this inexpensive old drug passes the BBB, it would be useful for therapy of formidable brain tumors such as glioma and NF2 tumors.

3.3.9 Salidroside

Extracts of plant adaptogens such as *Rhodiola rosea* (golden root) increase stress resistance in several models. These short plants grow on highlands in the Far East over 3000 m above sea level, such as Tibet and the Japanese Alps, and reddish root extracts have been used as an old Chinese traditional medicine. The major effective ingredient is a sugar called salidroside. In 2008, a Chinese group found that salidroside activates the tumor-suppressive kinase AMPK [71]. Shortly thereafter, a Dutch group led by Fred Wiegant at Utrecht University discovered that extracts of *R. rosea* (10–25 μg/ml) can extend the lifespan of *C. elegans* by activating the tumor-suppressive transcription factor FOXO, which eventually activates the Hsp16 gene [72]. Since AMPK is known to activate FOXO, it is most likely that the major life-extending ingredient in this extract is salidroside. Interestingly, however, a Polish group in Warsaw found that salidroside blocks tumor-induced angiogenesis, which is known to require both PAK1 and AMPK [73]. Since AMPK activation by salidroside alone could not block angiogenesis, this finding clearly indicates that salidroside must inactivate PAK1 as well as activate AMPK, just like more than a dozen natural PAK1-blocking compounds such as CAPE, curcumin, berberine, emodin, apigenin, resveratrol, and capsaicin/capsiate, as well as synthetic PAK1 blockers such as metformin, OSU-03012, and GW2974 (Table 3.1).

3.4   PAK1 Blockers = AMPK Activators

AMPK is an AMP-activated kinase that is activated when the AMP/ATP ratio increases. This ratio rises when the cellular glucose/ATP level drops, due to either calorie restriction (CR), fast or extensive physical exercise. Thus, AMPK serves as a sensor of the cellular glucose/ATP level and is activated in an AMP-dependent

Table 3.1 Anti-PAK1 and AMPK Activating Activity of Anti-PAK1 Products

Products	PAK1 Inactivation	AMPK Activation	Refs.
CAPE	+	+	[39,57]
Caffeic acid	+	+	[39,74]
Curcumin	+	+	[56,58]
Resveratrol	+	+	[75,76]
Berberine	+	+	[65,66]
Salidroside	+	+	[71,73]
Metformin	+	+	[77,78]
Emodin	+	+	[63,79]
Capsaicin	+	+	[80,81]
Apigenin	+	+	[82,83]
Rhizochalin	+	+	[84]
Honokiol	+	+	[85,86]
OSU-03012	+	+	[87,88]
GW2974	+	+	[89]
AG 879	+	+	[6,18]
AICAR	?	+	[90,91]
UnPAK309	+	−	[22]

manner. In most cases, these PAK1 blockers activate AMPK through another tumor-suppressing kinase called LKB1 that phosphorylates Thr 172 of AMPK [92]. AMPK phosphorylates several distinct proteins. As mentioned earlier, one of them is the glucose transporter protein GLUT-4, which is essential for the cellular uptake of glucose from the bloodstream. AMPK activates GLUT-4, leading to a transient rise of the cellular glucose level by lowering the blood glucose level. Another target of AMPK is the tumor-suppressing transcription factor FOXO, which is essential for longevity [93]. In mammals, AMPK activates FOXO, while PAK1 inactivates FOXO [94,95]. Thus, most of the PAK1 blockers, such as CAPE and curcumin, activate FOXO through these two distinct routes in a concerted manner. As discussed in detail in Chapter 7, at least three PAK1 blockers = AMPK activators (CA, salidroside, and curcumin) have been shown to extend the lifespan of the nematode *C. elegans* or the fruit fly *Drosophila* [72,96,97]. Among the major target genes of FOXO is HSP16, which encodes a small heat-shock protein, and this FOXO-HSP16 signaling pathway extends the lifespan of this worm by 50% [98]. We found that two PAK1 blockers from propolis, CAPE and ARC, strongly activate the HSP16 gene in *C. elegans* shortly after the heat shock and make this worm heat resistant, as does the PAK1-deficient mutation [41]. Heat-shock proteins in general protect essential cellular proteins from denaturation caused by heat [99]. Thus, it is most likely that both CAPE- and ARC-based propolis extracts would extend the lifespan of this worm, and probably of all human beings as well, including cancer/NF patients, by activating the longevity (FOXO-HSP16) signaling pathway. The stronger this signaling is, the longer we live.

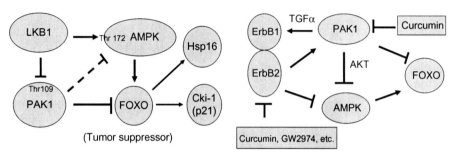

Figure 3.10 LKB1 controls both AMPK and PAK1.

Figure 3.11 ErbB2 controls PAK1 and AMPK.

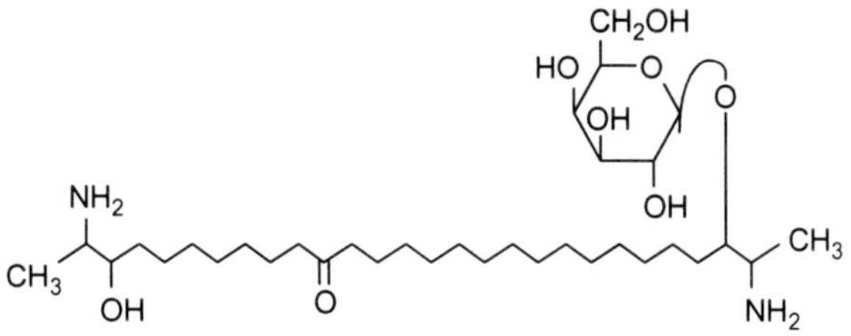

Figure 3.12 Rhizochalin (glucoside).

Why do most PAK1 blockers turn out to be AMPK activators, or vice versa? One of the mechanisms underlying this mysterious equation, PAK1 blockers = AMPK activators, is that LKB1 inactivates PAK1 by phosphorylating Thr 109, as well as simultaneously activating AMPK [77] (Figure 3.10). A second mechanism is based on the vicious oncogenic cycle that PAK1 and the Tyr kinase ErbB2 form: PAK1 activates another Tyr kinase called ErbB1 by upregulating its ligands that activate ErbB1, such as TGFα, and in turn ErbB1 activates ErbB2 by forming a heterodimer [7,100]. Both ErbB1 and ErbB2 are essential for the full activation of PAK1 [7]. Furthermore, ErbB2 somehow inactivates AMPK [89]. Thus, if either PAK1 or ErbB2 were blocked, this oncogenic cycle would stop, and AMPK would be reactivated (Figure 3.11).

In this context, it should be worth noting that aglycon of rhizochalin (Figure 3.12), a unique two-headed sphingolipid, was recently reported by a Korean group to activate AMPK and inactivate ERK and AP-1, both of which act downstream of PAK1 and suppress the anchorage-independent growth of human colon cancer cells with the IC_{50} below microM [84]. Rhizochalin was originally discovered in a Madagascar marine sponge called *Rhizochalina incrustata* in the late 1980s by Tatyana Makarieva's group at the Russian Academy of Sciences. Among the metabolites of rhizochalin, the aglycon was found to be the most potent suppressor of EGF-induced malignant transformation [101]. Since PAK1 is downstream of the EGF receptor ErbB1 and upstream of the kinase ERK, it is most likely that aglycon of rhizochalin blocks PAK1, as well as activates AMPK, just like other AMPK activators =PAK1 blockers. Like FTY720, sphingolipids in general activate PAK1 by binding around its GTPase-binding domain [102]. However, this two-headed sphingolipid appears to act as a rather potent antagonist of these PAK1-activating sphingolipids.

Acknowledgments

We are very grateful to the following funding sources for their generous support of our cancer/NF research: Deutsche Forschungs Gemeinschaft (DFG), Denise Terrill Classic Award/ Texas NF Foundation, NF Cure Japan, Peterson Foundation (Philadelphia), Yamada Bee Farm, and Neurofibromatosis Inc. Also we are indebted to the following colleagues: Drs. Tony Burgess, Hong He, Yumiko Hirokawa, Maria Demestre, Tamotsu Sudo, Toshiro Ohta, Hitoshi Hori, Ken Hashimoto, Hidenori Nakajima, John Beutler, John Wood, Lan Kluwe, and Victor Mautner for their great teamwork during the past decade.

References

[1] Maruta H, Burgess AW. Regulation of the Ras signalling network. Bioessays 1994;16:489–96.

[2] Tang Y, Chen Z, Ambrose D, Liu J, et al. Kinase-deficient Pak1 mutants inhibit Ras transformation of Rat-1 fibroblasts. Mol Cell Biol 1997;17:4454–64.

[3] Manser E, Loo T, Koh C, Zhao Z, et al. PAK kinases are directly coupled to the PIX family of nucleotide exchange factors. Mol Cell 1998;1:183–92.

[4] Obermeier A, Ahmed S, Manser E, Ye SC, et al. PAK promotes morphological changes by acting upstream of Rac. EMBO J 1998;17:4328–39.

[5] Nur-E-Kamal MSA, Kamal JM, Quresh MM, Maruta H. The CDC42-specific inhibitor derived from ACK blocks the v-Ha-RAS-induced transformation. Oncogene 1999;18:7787–93.

[6] He H, Hirokawa Y, Manser E, Lim L, et al. Signal therapy for RAS induced cancers in combination of AG 879 and PP1, specific inhibitors for ErbB2 and Src family kinases, that block PAK activation. Cancer J 2001;7:191–202.

[7] Zhao L, Ma QL, Calon F, Harris-White M, et al. Role of PAK pathway defects in the cognitive deficits of Alzheimer's disease. Nat Neurosci 2006;9:234–42.

[8] Viaud J, Peterson J. An allosteric kinase inhibitor binds the p21-activated kinase autoregulatory domain covalently. Mol Cancer Ther 2009;8:2559–65.

[9] Maroney A, Glicksman M, Basma A, Walton K, et al. Motoneuron apoptosis is blocked by CEP-1347 (KT 7515), a novel inhibitor of the JNK signaling pathway. J Neurosci 1998;18:104–11.

[10] Nheu T, He H, Hirokawa Y, Tamaki K, et al. The K252a derivatives, inhibitors for the PAK/MLK kinase family selectively block the growth of RAS transformants. Cancer J 2002;8:328–36.

[11] Lei M, Lu W, Meng W, et al. Structure of PAK1 in an auto-inhibited conformation reveals a multistage activation switch. Cell 2000;102:387–97.

[12] Hirokawa Y, Arnold M, Nakajima H, Zalcberg J, et al. Signal therapy of breast cancer xenograft in mice by the HDAC inhibitor FK228 that blocks the activation of PAK1 and abrogates the tamoxifen-resistance. Cancer Biol Ther 2005;4:956–60.

[13] Schupp P, Steube K, Meyer C, et al. Anti-proliferative effects of a new staurosporine derivative isolated from a marine ascidian and its predatory flat worm. Cancer Lett 2001;174:165–72.

[14] Onaka H, Taniguchi S, Igarashi Y, Furumai T. Cloning of the staurosporine biosynthetic gene cluster from Streptomyces, sp. TP-ao274 and its heterologous expression in *Streptomyces lividans*. J Antibiot (Tokyo) 2002;55:1063–71.

[15] Tipper J, Citron B, Ribeiro P, Kaufmann S. Cloning and expression of rabbit and human tryptophan hydroxylase cDNA in *E. coli*. Arch Biochem Biophys 1994;315:445–53.

[16] He H, Hirokawa Y, Levitzki A, Maruta H. An anti-Ras cancer potential of PP1, an inhibitor specific for Src family kinases: *in vitro* and *in vivo* studies. Cancer J 2000;6:243–8.

[17] Wu W, Sun Z, Wu J, Peng X, et al. Trihydrophobin 1 phosphorylation by c-Src regulates MAPK/ERK signaling and cell migration. PLoS One 2012;7:e29920.

[18] He H, Hirokawa Y, Gazit A, Yamashita Y, et al. The Tyr kinase inhibitor AG879, that blocks the ETK–PAK1 interaction, suppresses the RAS-induced PAK1 activation and malignant transformation. Cancer Biol Ther 2004;3:96–101.

[19] Hirokawa Y, Levitzki A, Lessene G, Baell J, et al. Signal therapy of human pancreatic cancer and NF1-deficient breast cancer xenograft in mice by a combination of PP1 and GL-2003, anti-PAK1 drugs (Tyr kinase inhibitors). Cancer Lett 2007;245:242–51.

[20] Schindler T, Sicheri F, Pico A, Gazit A, Levitzki A, Kuriyan J. Crystal structure of Hck in complex with a Src family-selective tyrosine kinase inhibitor. Mol Cell 1999;3:639–48.

[21] Yadav V, Denning MF. Fyn is induced by Ras/PI3K/Akt signaling and is required for enhanced invasion/migration. Mol Carcinog 2011;50:346–52.

[22] Murray B, Guo CX, Piraino J, Westwick J, et al. Small-molecule PAK inhibitor PF-3758309 is a potent inhibitor of oncogenic signaling and tumor growth. Proc Natl Cad Sci USA 2010;107:9446–51.

[23] Qu J, Li X, Novitch B, Zheng Y, et al. PAK4 kinase is essential for embryonic viability and for proper neuronal development. Mol Cell Biol 2003;23:7122–33.

[24] Ueda H, Nakajima H, Hori Y, Fujita T, et al. FK228, a novel antitumor bicyclic depsipeptide, I. Taxonomy, fermentation, isolation, physicochemical and biological properties, and antitumor activity. J Antibiot 1994;47:301–10.

[25] Nakajima H, Kim YB, Terano H, Yoshida M, et al. FK228, a potent anti-tumor antibiotic, is a novel histone deacetylase inhibitor. Exp Cell Res 1998;241:126–33.

[26] Kobayashi Y, Ohtsuki M, Murakami T, Kobayashi T, et al. Histone deacetylase inhibitor FK228 suppresses the Ras-MAP kinase signaling pathway by upregulating Rap1 and induces apoptosis in malignant melanoma. Oncogene 2006;25:512–24.

[27] Kodani M, Igishi T, Matsumoto S, Chikumi H, et al. Suppression of PI 3-kinase/AKT signaling pathway is a determinant of the sensitivity to a novel HDAC inhibitor, FK228, in lung adenocarcinoma cells. Oncol Rep 2005;13:477–83.

[28] Campas-Moya C. Romidepsin (FK228) for the treatment of CTCL. Drugs Today 2009;45:787.

[29] Tang Y, Marwaha S, Rutkowski J, Tennekoon G, et al. A role for Pak protein kinases in Schwann cell transformation. Proc Natl Acad Sci USA 1998;95:5139–44.

[30] Hirokawa Y, Tikoo A, Huynh J, Utermark T, et al. A clue to the therapy of neurofibromatosis type 2: NF2/merlin is a PAK1 inhibitor. Cancer J 2004;10:20–6.

[31] Nur-E-Kmal MSA, Varga M, Maruta H. The GTPase-activating NF1 fragment of 91 amino acids reverses the v-Ha-RAS-induced malignant phenotype. J Biol Chem 1993;268:22331–22337.

[32] Tikoo A, Varga M, Ramesh V, Gusella J, et al. An anti-RAS function of neurofibromatosis type 2 gene product (NF2/Merlin). J Biol Chem 1994;269:23387–23390.

[33] Kiosses W, Hood J, Yang S, Gerritsen M, et al. A dominant-negative p65 PAK peptide inhibits angiogenesis. Circ Res 2002;90:697–702.

[34] Allen J, Jaffer Z, Park S, Burgin S, et al. PAK1 regulates mast cell degranulation via effects on calcium mobilization and cytoskeletal dynamics. Blood 2009;113:2695–705.

[35] Yoshioka K, Foletta V, Bernard O, Itoh K. A role for LIM kinase in cancer invasion. Proc Natl Acad Sci USA 2003;100:7247–52.

[36] Grunberger D, Banerjee R, Eisinger K, Oltz E, et al. Preferential cytotoxicity on tumor cells by caffeic acid phenethyl ester isolated from propolis. Experientia 1988;44:230–2.

[37] Matsuno T, Jung SK, Matsumoto Y, Saito M, et al. Preferential cytotoxicity to tumor cells of artepillin C isolated from propolis. Anticancer Res 1997;17:3565–8.

[38] Awale S, Li F, Onozuka H, Esumi H, et al. Constituents of Brazilian red propolis and their preferential cytotoxic activity against human pancreatic PANC-1 cancer cell line in nutrient-deprived condition. Bioorg Med Chem 2008;16:181–9.

[39] Xu JW, Ikeda K, Kobayakawa A, Ikami T, et al. Down-regulation of Rac1 activation by caffeic acid in aortic smooth muscle cells. Life Sci 2005;76:2861–72.

[40] Messerli S, Ahn MR, Kunimasa K, Yanagihara M, et al. Artepillin C (ARC) in Brazilian green propolis selectively blocks the oncogenic PAK1 signaling and suppresses the growth of NF tumors in mice. Phytother Res 2009;23:423–7.

[41] Maruta H. An innovated approach to in vivo screening for the major anti-cancer drugs Horizons in cancer research, 41. : Nova Science Publishers; 2010. pp. 249–59.

[42] Demestre M, Messerli S, Celli N, Shahhossini M, et al. CAPE (caffeic acid phenethyl ester)-based propolis extract (Bio 30) suppresses the growth of human neurofibromatosis (NF) tumor xenografts in mice. Phytother Res 2009;23:226–30.

[43] Sicard A, Semblat JP, Doerig C, Hamelin R, et al. Activation of PAK–MEK signaling pathway in malaria parasite-infected erythrocytes. Cell Microbiol 2011;13:836–45.

[44] Wang JX, Tang W, Shi LP, Wan J, et al. Investigation of the immunosuppressive activity of artemether on T-cell activation and proliferation. Brit J Pharmacol 2007;150:652–61.

[45] Du JH, Zhang HD, Ma ZJ, Ji KM. Artesunate induces oncosis-like cell death in vitro and has anti-tumor activity against pancreatic cancer xenografts in vivo. Cancer Chemother Pharmacol 2010;65:895–902.

[46] Trager W, Polonsky J. The antimalaria activity of Quassinoids against chloroquine-resistant P. falciparum in vitro. Am J Trop Med Hyg 1981;30:531–7.

[47] Ghosh P, Larrahond J, LeQuesne P, Raffauf R. Antitumor plants. IV. Constituents of S. versicolor. Lloydia 1977;40:364–9.

[48] Valeriote F, Corbett T, Grieco P, Moher E, et al. Anticancer activity of glaucarubinone analogues. Oncol Res 1998;10:201–8.

[49] Beutler J, Kang MI, Robert F, Clement J, et al. Quassinoid inhibition of AP-1 function does not correlate with cytotoxicity or protein synthesis inhibition. J Nat Prod 2009;72:503–6.

[50] Adam L, Vadlamudi R, Mandal M, Chernoff J, et al. Regulation of microfilament organization and invasiveness of breast cancer cells by kinase dead PAK1. J Biol Chem 2000;275:12041–12050.

[51] Mendora-Gamboa E, Siwak D, Tari A. The HER2/Grb2/Akt pathway regulates DNA binding activity of AP-1 in breast cancer cells. Oncol Rep 2004;12:904–8.

[52] Zarse K, Bossecker A, Mueller-Kuhrt L, Siems K, et al. The phytochemical glaucarubinone promotes mitochondrial metabolism, reduces body fat and extends the life span of C. elegans. Horm Metab Res 2011;43:241–3.

[53] Castelletti D, Fiaschetti G, Di Dato V, Ziegler U, et al. The quassinoid derivative NBT-272 targets both the AKT and ERK signaling pathways in embryonal tumors. Mol Cancer Ther 2010;9:3145–57.

[54] Yang SH, Wang X, Contino G, Leasa M, et al. Pancreatic cancers require autophagy for their growth. Gene Develop 2011;25:717–29.

[55] Hirokawa Y, Nheu T, Grimm K, Mautner V, et al. Sichuan pepper extracts block the PAK1/cyclin D1 pathway and the growth of NF1-deficient cancer xenograft in mice. Cancer Biol Ther 2006;5:305–9.

[56] Cai XZ, Wang J, Li XD, Wang GL, et al. Curcumin suppresses proliferation and invasion in human gastric cancer cells by down-regulation of PAK1 activity and cyclin D1 expression. Cancer Biol Ther 2009;8:1360–8.

[57] Lee ES, Uhm KO, Lee YM, Han M, et al. CAPE stimulates glucose uptake through AMPK activation in skeletal muscle cells. Biochem Biophys Res Commun 2007;361:854–8.

[58] Pan W, Yang C, Cao C, Song X, et al. AMPK mediates curcumin-induced cell death of CaOV3 ovarian cancer cells. Oncol Rep 2008;20:1553–9.

[59] Li L, Braiteh FS, Kurzrock R. Liposome-encapsulated curcumin: *in vitro* and *in vivo* effects on proliferation, apoptosis, signaling, and angiogenesis. Cancer 2005;104:1322–31.

[60] Mach C, Mathew L, Mosley S, Kurzrock R, et al. Determination of minimum effective dose and optimal dosing schedule for liposomal curcumin in a xenograft human pancreatic cancer model. Anticancer Res 2009;29:1895–9.

[61] Yadav V, Suresh S, Devi K, Yadav S. Effect of cyclodextrin complexation of curcumin on its solubility and anti-angiogenic and anti-inflammatory activity in Rat colitis model. AAPS Pharm Sci Tech 2009;10:752–62.

[62] Zhang L, Lao YK, Xia W, Hortbagyi G, et al. Tyr kinase inhibitor emodin suppresses the growth ErbB2-overexpressing breast cancer cells in athymic mice. Clin Cancer Res 1999;5:343–53.

[63] Xue J, Ding W, Liu Y. Anti-diabetic effects of emodin involved in the activation of PPARgamma in high-fat diet-fed and low dose of streptozotocin-induced diabetic mice. Fitoterapia 2010;81:173–7.

[64] Chang KS, Gao C, Wang LC. Berberine-induced morphologic differentiation and down-regulation of c-Ki-ras2 protooncogene expression in human teratocarcinoma cells. Cancer Lett 1990;55:103–8.

[65] Brusq JM, Ancellin N, Grondin P, Guillard R, et al. Inhibition of lipid synthesis through activation of AMPK: an additional mechanism for the hypolipidemic effects of berberine. J Lipid Res 2006;47:1281–8.

[66] Tsang CM, Lau EP, Di K, Cheung PY, et al. Berberine inhibits Rho GTPases and cell migration at low doses but induces G2 arrest and apoptosis at high doses in human cancer cells. Int J Mol Med 2009;24:131–8.

[67] Bahar M, Deng Y, Zhu XH, He SS, et al. Potent anti-protozoan activity of a novel semisynthetic berberine derivative. Bioorg Med Chem Lett 2011;21:2606–10.

[68] Omura S, Crump A. The life and times of ivermectin: a success story. Nat Rev Microbiol 2004;2:894–9.

[69] Hashimoto H, Messerli S, Sudo T, Maruta H. Ivermectin inactivates the kinase PAK1 and block the PAK1-dependent growth of human ovarian cancer and NF2 tumor cell lines. Drug Discov Ther 2009;3:243–6.

[70] Drinyaev V, Mosin V, Kluglyak E, Novik. T, et al. Antitumor effect of avermectin. Eur J Pharmacol 2004;501:19–23.

[71] Li HB, Ge YK, Zheng XX, Zhang L. Salidroside stimulated glucose uptake in skeletal muscle cells by activating AMPK. Eur J Pharmacol 2008;588:165–9.

[72] Wiegant F, Surinova S, Ytsma E, Langelaar-Makkinje M, et al. Plant adaptogens increase lifespan and stress resistance in *C. elegans*. Biogerontology 2009;10:27–42.

[73] Skopin'ska-Rozewska E, Malinowski M, Wasiutyn'ski A, Sommer E, et al. The influence of *Rhodiola quadrifida* 50% hydro-alcoholic extract and salidroside on tumor-induced angiogenesis in mice. Pol J Vet Sci 2008;11:97–104.

[74] Tsuda S, Egawa T, Ma X, Oshima R, et al. Coffee polyphenol caffeic acid but not chlorogenic acid increases AMPK and insulin-independent glucose transport in rat skeletal muscle. J Nutr Biochem 2012;23:1403–9.

[75] Waite KA, Sinden MR, Eng C. Phytoestrogen exposure elevates PTEN levels. Hum Mol Genet 2005;14:1457–63.

[76] Penumathsa SV, Thirunavukkarasu M, Zhan L, Maulik G, et al. Resveratrol enhances GLUT-4 translocation to the caveolar lipid raft fractions through AMPK/Akt/eNOS signalling pathway in diabetic myocardium. J Cell Mol Med 2008;12:2350–61.

[77] Deguchi A, Miyashi H, Kojima Y, Okawa K, et al. LKB1 inactivates PAK1 by phosphorylation of Thr 109 in the p21-binding domain. J Biol Chem 2010;285:18283–18290.

[78] Zhou GC, Meyers R, Li Y, Chen Y, et al. Role of AMPK in mechanism of metformin. J Clin Invest 2001;108:1167–74.

[79] Huang Q, Shen HM, Ong CN. Emodin inhibits tumor cell migration through suppression of the PI-3 kinase/CDC42/Rac pathway. Cell Mol Life Sci 2005;62:1167–75.

[80] Shin DH, Kim OK, Jun HS, Kang MK. Inhibitory effect of capsaicin on B16-F10 melanoma cell migration via the PI-3 kinase/Akt/Rac1 signal pathway. Exp Mol Med 2008;40:486–94.

[81] Hwang JT, Park IJ, Shin JI, Lee YK, et al. Genistein, EGCG, and capsaicin inhibit adipocyte differentiation process via activating AMPK. Biochem Biophys Res Commun 2005;338:694–9.

[82] Way TD, Kao MC, Lin JK. Apigenin induces apoptosis through proteosomal degradation of ErBB2 in ErBB2-overexpressing breast cancer cells via the PI3-kinase/AKT-dependent pathway. J Biol Chem 2004;279:4479–89.

[83] Zang M, Xu S, Maitland-Toolan K, Zuccollo A, et al. Polyphenols stimulate AMPK, lower lipids, and inhibit accelerated atherosclerosis in diabetic LDLR-deficient mice. Diabetes 2006;55:2180–91.

[84] Khanal P, Kang BS, Yun HJ, Cho HG, et al. Aglycon of rhizochalin from the *Rhizochalina incrustata* induces apoptosis via activation of AMPK in HT-29 colon cancer cells. Biol Pharm Bull 2011;34:1553–9.

[85] Bai X, Cerimele F, Ushio-Fukai M, Waqas M, et al. Honokiol, a small molecular weight natural product, inhibits angiogenesis *in vitro* and tumor growth *in vivo*. J Biol Chem 2003;278:35501–35507.

[86] Harada S, Kishimoto M, Kobayashi M, Nakamoto K, et al. Honokiol suppresses the development of post-ischemic glucose intolerance and neuronal damage in mice. J Nat Med 2012;66:591–9.

[87] Porchia L, Guerra M, Wang YC, Zhang YL, et al. OSU-03012, a celecoxib derivative, directly targets PAK. Mol Pharmacol 2007;72:1124–31.

[88] Bai LY, Weng JR, Tsai CH, Sargeant A, et al. OSU-03012 sensitizes TIB-196 myeloma cells to imanitib mesylate via AMPK and STAT3 pathways. Leuk Res 2010;34:826–30.

[89] Spector N, Yarden Y, Smith B, Lyass L, et al. Activation of AMPK by ErbB2/ErbB1 Tyr kinase inhibitor protects cardiac cells. Proc Natl Acad Sci USA 2007;104:10607–10612.

[90] Sullivan J, Brocklehurst K, Marley A, Carey K, et al. Inhibition of lipolysis and lipogenesis in isolated rat adipocytes with AICAR, a cell-permeable activator of AMPK. FEBBS Lett 1994;353:33–6.

[91] Sun Y, Connors K, Yang DQ. AICAR induces phosphorylation of AMPK in an ATM-dependent, LKB1-independent manner. Mol Cell Biochem 2007;306:239–45.

[92] Shaw R, Kosmatka S, Bardeesy N, Hurley R, et al. The tumor suppressor LKB1 kinase directly activates AMPK and regulates apoptosis in response to energy stress. Proc Natl Acad Sci USA 2004;101:3329–35.

[93] Kenyon C, Chang J, Gensch E, Rudner A, et al. A C. elegans mutant that lives twice as long as wild type. Nature 1993;366:461–4.

[94] Greer E, Dowlatshahi D, Banko M, Villen J, et al. An AMPK–FOXO pathway mediates longevity induced by a novel method of dietary restriction in C. elegans. Curr Biol 2007;17:1646–56.

[95] Vadlamudi R, Kumar R. p21-Activated kinase 1 (PAK1): an emerging therapeutic target. Cancer Treat Res 2004;119:77–88.

[96] Lee KS, Lee BS, Semnani S, Avanesian A, et al. Curcumin extends life span, improves health span, and modulates the expression of age-associated aging genes in D. melanogaster. Rejuven Res 2010;13:561–70.

[97] Pietsch K, Saul N, Shumon Chakrabarti S, Stürzenbaum S, et al. Hormetins, antioxidants and prooxidants: defining quercetin-, caffeic acid- and rosmarinic acid-mediated life extension in C. elegans. Biogerontology 2011;12:329–47.

[98] Walker G, White T, McColl G, Jenkins N, et al. Heat shock protein accumulation is upregulated in a long-lived mutant of C. elegans. J Gerontol A Biol Sci Med Sci 2001;56:B281–7.

[99] Engelberg D, Zandi E, Parker C, Karin M. The yeast and mammalian RAS pathways control transcription of heat shock genes independently of heat shock transcription factor. Mol Cell Biol 1994;14:4929–37.

[100] Goldman R, Levy R, Peles E, Yarden Y. The heterodimerization of the ErbB1 and ErbB2 receptors in human breast carcinoma cells: a mechanism for receptor transregulation. Biochemistry 1990;29:11024–11028.

[101] Fedrov S, Makarieva T, Guzii A, Shubina L, et al. Marine two-headed sphingolipid-like compound rhizochalin inhibits EGF-induced transformation of JB6 P+ Cl41 cells. Lipids 2009;44:777–85.

[102] Bokoch G, Reilly A, Daniels H, King C, et al. A GTPase-independent mechanism of PAK1 activation. Regulation by sphingosine and other biologically active lipids. J Biol Chem 1998;273:8137–44.

4 PAK1–3 in Infectious Diseases

Hiroshi Maruta

NF/TSC Cure Organisation, Melbourne, Australia

4.1 Malaria Infection

Malaria still remains one of the most devastating infectious diseases, especially in sub-Saharan Africa, where it claims the lives of more than 1 million persons every year, most of whom are young children. The agent responsible for the most severe form of human malaria is *Plasmodium falciparum*, an intracellular parasite belonging to the phylum Apicomplexa. Transmitted to the human host through the bite of an infected Anopheles mosquito, the parasite quickly reaches the liver and invades a hepatocyte, where a first round of asymptomatic asexual multiplication (exo-erythrocytic schizogony) requires seven days to reach completion. In order to maintain survival of its host hepatocyte for that duration, the parasite activates the NF-κB pathway, thereby preventing the apoptosis of its host, among others [1,2].

Almost a decade ago, we started suspecting that malaria infection might require one of the group 1 PAKs (PAK1–3) because an old antimalarial drug called quinidine blocks PAK1-dependent growth of RAS-transformed cells with IC_{50} around 25 μM [3]. However, quinidine is rather toxic to be used as an anticancer drug, and we began to identify a far more potent and safer alternative. One of them was FK228, which eventually inactivates PAK1 with IC_{50} below 1 nM through inhibition of histone deacetylase (HDAC) and suppresses the growth of a variety of PAK1-dependent solid tumors such as pancreatic and breast cancers, as well as mesotheliomas, schwannoma, and malignant peripheral nerve sheath tumor (MPNST), as discussed in Chapters 3 and 5 [4].

One of the urgent tasks in malaria therapy is to develop or identify a new drug that is effective on malaria strains that have become resistant to widely used antimalarial drugs such as chloroquine. In collaboration with Alan Cowman's group at the Walter and Eliza Hall Institute (WEHI) in Melbourne, we found that FK228 indeed blocks chloroquine-resistant malaria infection, at least in cell culture (A Cowman and H Maruta, unpublished observation), strongly suggesting, if not yet proving, that one of the PAK1–3 kinases is essential for malaria infection.

A few years later, a kinome-wide RNAi study identified five host protein kinases implicated in hepatocyte infection by *Plasmodium*, in particular revealing that PAK3 inhibition leads to a reduction in parasite infection [5] and clearly indicating that at least PAK3 is involved in malaria infection.

Following hepatocyte rupture, free malaria parasites (merozoites) invade red blood cells and initiate cycles of asexual multiplication (erythrocytic schizogony)

PAKs, RAC/CDC42 (p21)-activated Kinases. DOI: http://dx.doi.org/10.1016/B978-0-12-407198-8.00004-7

that are responsible for malaria pathogenesis. Each cycle consists of a succession of developmental stages, starting with the so-called ring stage that immediately follows invasion (named after the shape of the intracellular parasite in Giemsa-stained blood smears), followed by the trophozoite stage, during which the parasite grows by feeding, largely on host cell hemoglobin, and finally the schizont stage, in which a multinucleated parasite results from multiple nuclear divisions prior to cytokinesis and merozoite release. During the invasion process, a parasitophorous vacuole (PV) is formed, inside which the parasite will reside during its development in the red blood cell. Maturation of the parasite in the erythrocyte results in an important reorganization of the red blood cell cytoskeleton [6], and the infected erythrocyte becomes spherical and rigid. Such a red cell would be eliminated by the spleen if the parasite had not developed a strategy to adhere to endothelial cells and be sequestered from circulation, thus avoiding the spleen [7]. During parasite maturation, numerous *Plasmodium* proteins are exported to the erythrocyte cytosol and to the plasma membrane through a specific translocon located at the PV membrane [8]; these include the variant antigen of the PfEMP1 family that mediates cytoadhesion [9]. Conversely, it has recently been shown that the parasite imports a host cell protein (peroxiredoxin II) [10], and this may occur for many other host cell proteins. Such protein trafficking is indicative of an intimate molecular interaction between parasite and host cell.

Moreover, host erythrocyte signaling is essential for malaria parasite entry and maturation. Indeed, blocking G protein-coupled receptors (GPCR) and heterotrimeric G protein signaling inhibits *Plasmodium* entry and development in the red blood cell [11,12]. However, an extensive investigation of host erythrocyte signaling events triggered during *Plasmodium* proliferation has been made difficult by the impossibility of performing reverse genetics and RNAi approaches in a mature red blood cell.

While working recently in Lausanne, Switzerland, Christian Doerig's group (currently at Monash University in Melbourne) found that several structurally distinct MEK inhibitors block parasite development at the trophozoite stage, and that under MEK inhibitor treatment, parasite DNA replication was severely impaired [13]. Initially they thought that the target was a parasite-encoded MEK; however, subsequent availability of the *P. falciparum* genome sequence [14] and *in silico* characterization of the parasite's kinome revealed the absence of genes-encoding MEK homologues in the plasmodial genome [15,16].

Thus, it is most likely that the inhibitors exerted their effect through inhibition of a host erythrocyte MEK. Later it was revealed that human MEK1 is phosphorylated on Ser 298 in infected erythrocytes, but not (or to very low levels) in uninfected erythrocytes. Ser 298 phosphorylation is known to promote MEK1 activation, and PAK is the only kinase determined so far to phosphorylate MEK on this residue. As expected, phosphorylation of PAK on Ser141 was higher in *P. falciparum*-infected erythrocytes than in uninfected erythrocytes. Furthermore, incubation of *P. falciparum* cultures with IPA-3, a PAK1–3 inhibitor [17], blocked parasite maturation and multiplication as well as reducing MEK phosphorylation at Ser 298 [13]. Taken together, these findings indicate that a host erythrocyte PAK–MEK pathway is activated by *P. falciparum* infection and is required for its further propagation.

Accordingly, two very intriguing hypotheses emerged: (1) PAK1–3 blockers could be used as effective new therapeutics for malaria in the future and (2) some older antimalarial drugs may block the activation of PAK1–3. As discussed earlier, FK228, the most potent HDAC inhibitor (and PAK1 blocker), which was recently marketed under the brand name Istodax by Celgene only for cutaneous T-cell lymphoma (CTCL) therapy, indeed blocks chloroquine-resistant malaria infection in cell culture. Regarding the latter hypothesis, in addition to quinidine, several other antimalarial drugs are now known to block PAK1 activation: artemisinin, chloroquine, curcumin, resveratrol, glaucarubinone, berberine, and propolis. For details, see Chapter 3.

4.2 Enterobacterial Infection

4.2.1 Helicobacter pylori

Helicobacter pylori is a Gram-negative, microaerophilic bacterium found in the stomach, where it adheres to the surface of epithelial cells. More than 50% of the world's population harbors *H. pylori* in their upper gastrointestinal tract. However, over 80% of infected individuals show no symptoms. *H. pylori* bacteria are present in patients with chronic gastritis, gastric ulcers, and gastric carcinoma [18,19].

Between 50% and 60% of *H. pylori* isolates possess a DNA segment called the CagA pathogenicity island (PAI; reviewed in Ref. [20]). Patients infected with strains carrying the Cag PAI display a stronger inflammatory response in the stomach and are at a greater risk of developing peptic ulcers or stomach cancer than those infected with strains lacking the island [21]; infection with CagA-positive *H. pylori* is the strongest risk factor for gastric carcinoma [22].

H. pylori infection leads to a profound reorganization of the host cell cytoskeleton, with actin polymerization and membrane ruffling appearing at the cell periphery [23]. After CagA-positive *H. pylori* adheres to gastric epithelial cells, the CagA protein is translocated into gastric epithelial cells through a type IV secretion system [24]. Inside the cell, CagA is Tyr phosphorylated and stimulates several distinct signaling pathways [25].

One of its effects is activation of the GTPases RAC and CDC42, leading to the activation of PAK1 [26]. A Korean group recently found that PAK-interactive exchange factor (PIX), an SH3 adaptor protein, plays a role in the CagA-induced activation of PAK1 [23]. Phosphorylated CagA leads to dephosphorylation of PIX and may thereby modulate cytoskeletal changes in gastric epithelial cells through PAK regulation [27]. Moreover, downregulation of PIX through a siRNA approach blocks PAK1 activation by CagA [27]. Additionally, PIX siRNA suppressed IL-8 induction after translocation of CagA into the cells, indicating that IL-8 expression depends on the CagA–PIX interaction [27]. *H. pylori* infection leads to an important pro-inflammatory response, induced mainly by the interleukin-1 beta (IL-1β) cytokine. *H. pylori* lipopolysaccharide (LPS) induced direct interaction between PAK1 and caspase-1 (a protease that is required for the maturation of pro-IL-1β into the active cytokine), and in turn PAK1 phosphorylates caspase-1.

These data indicate that upon *H. pylori* LPS induction, activated PAK1 stimulates IL-1 production [28] and therefore plays a major role in the inflammatory response that is one of the hallmarks of *H. pylori* pathogenesis.

4.2.2   E. coli (Hemorrhagic Strain)

Neal Alto's group at University of Texas Medical Center (UTMC) in Dallas (Fig. 4.1) recently found that a virulent (hemorrhagic) strain of *E. coli* (O157: H7) uses a bacterial virulent protein called EspG, instead of RAC/CDC42, to activate PAK1–3 by binding to its AID to dissociate the inactive homodimer of PAK1–3 [29,30] (Figure 4.2B). This RAC/CDC42-independent activation of PAKs leads to disruption of gap junction/tight junction of bowel cells and eventually causes a severe infantile diarrhea, often leading to death [31]. This finding suggests the possibility that a chronic infection with this virulent strain of *E. coli* might cause a variety of cancers and other PAK1-dependent diseases as well.

Very interestingly, according to a previous study by Mike Rosen's group at UTMC (Fig. 4.3), another bacterial virulent protein, called EspFU, similarly binds the AID of N-WASP, a CDC42-activated Arp2/3-binding protein, to dissociate the VCA (α5 helix)-AID (α3 helix) interlocking in N-WASP molecules (Figure 4.2A), leading to the full activation of N-WASP [32]. The resulting N-WASP–Arp2/3 complex formation via VCA induces a rapid actin polymerization [33]. These two related exciting discoveries suggest that the ability of virulent bacterial proteins to identify

Figure 4.1 Dr. Neal Alto's Team at UTMC.

novel regulatory principles of host signaling enzymes highlights the anticipated mul-
tilevel nature of PAK/N-WASP activation and makes them effective tools to study
mammalian RAC/CDC42-dependent signaling pathways, as well as to screen for
synthetic chemicals or natural products that selectively block the potentially oncogenic PAK/N-WASP signaling pathways.

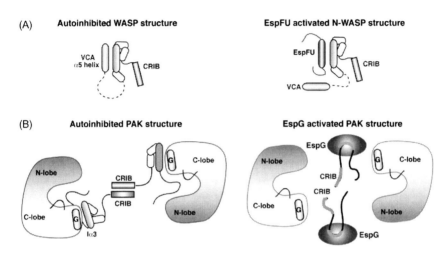

Figure 4.2 EspG-induced activation of PAKs, compared with EspFU-induced activation of
WASPs.
Source: The figure was kindly provided by Dr. Neal Alto.

Figure 4.3 Dr. Mike Rosen's Lab at UTMC.

4.3   Viral Infection

4.3.1   AIDS (HIV)

The United Nations Joint Programme on HIV/AIDS (UNAIDS) and the World Health Organization (WHO) estimated that acquired immune deficiency syndrome (AIDS) killed more than 25 million people between 1981 and 2005, making it one of the most destructive pandemics in recorded history. Around 30 million people are infected worldwide and there are three million new infections each year. AIDS is caused by the human immunodeficiency virus (HIV) retrovirus. HIV infects vital cells in the human immune system such as helper T cells (specifically CD4+ T cells), macrophages, and dendritic cells. When CD4+ T-cell numbers decline below a critical threshold of 200 cells per μl, cell-mediated immunity is lost, exposing the patient to a wide variety of opportunistic microbial infections. Thus, the major cause of death of AIDS patients is deadly infection by a variety of pathogenic bacteria, fungi, and tumor viruses, mainly due to the loss of their immune system.

Although treatments for HIV/AIDS can slow down the course of the disease, there is no known cure or HIV vaccine. Antiretroviral treatment reduces both HIV/AIDS deaths and new infections, but these drugs are expensive and the medications are not available in all countries, in particular underdeveloped countries in Africa and South Asia. Due to the difficulty of treating HIV infection, preventing infection is a key aim in controlling the AIDS pandemic, with health organizations promoting "safe sex" and "no needle-exchange" programs in attempts to slow the spread of this virus.

However, once people are infected with HIV, effective and relatively inexpensive therapeutics are needed for the treatment of HIV infection.

The negative factor (NEF) is a critical viral protein responsible for AIDS pathogenesis [34,35]. NEF is implicated in many aspect of the virus life cycle such as replication, spread, and immune evasion. In 1994, NEF was found to be associated with a Ser/Thr kinase (then called NAK for NEF-associated kinase) encoded by the host cell [35]. This kinase was later identified as PAK2 by Paul Luciw's group at the University of California, Davis, and a few other groups [36–38]. In 2001 Andreas Baur's group at Erlangen University in Germany found the activation of PAK by NEF via a phosphatidylinositol-3 kinase (PI-3-kinase) pathway, leading to the stimulation of phosphorylation of the Bcl2 antagonist of cell death (BAD) at Ser 112 [39]. As the phosphorylation of BAD at Ser 112 has an antiapoptotic effect, it is likely that the virus uses this strategy to block the apoptotic signals triggered by its presence in the cell. Interestingly, this result is consistent with the finding that NEF activates PAK by recruiting the kinase at the lipid raft fraction of the plasma membrane [40]. Indeed, it is known that BAD possesses two lipid-binding domains [41] and can be regulated through its attachment to lipid rafts [42–44]. Which member of the PAK family binds NEF is still the subject to debate. The original school of thought favored PAK2 as the NEF-binding PAK isoform [36–38]. In 2006, however, based on a siRNA approach, Deborah Nguyen's group at Novartis in San Diego demonstrated that PAK1, but not PAK2, inhibition strongly reduced HIV infection in multiple cell systems, establishing that PAK1 is the essential NEF partner during virus infection [45]. It is still possible that both PAK1 and PAK2 play a critical role in the HIV pathogenicity, with each protein having a distinct function.

4.3.2 Influenza Virus

A bee product called propolis, an alcohol extract of honeycomb resin, has been used as a traditional medicine for the therapy of a variety of inflammatory diseases such as asthma and arthritis and viral infections such as flu since the ancient Egyptian era. However, the molecular mechanism underlying the antiviral action of propolis remained totally unknown until recently. The first clue to our understanding of its antiviral mechanism was the discovery of caffeic acid (CA), caffeic acid phenethyl ester (CAPE), apigenin and artepillin C (ARC) in propolis [46–48]. As discussed in Chapter 3 in detail, these polyphenols share a unique common biological property: they block the oncogenic kinase called PAK1 [46–48].

Once PAK1 was implicated in the replication of several viruses, including HIV, as discussed above, a Korean group led by Young-Ki Choi first started testing whether replication of the influenza A virus stimulated the autophosphorylation of PAK1 and showed that this is indeed the case [49]. Moreover, transfection of a constitutively active form of PAK1 (rendered active through the phosphomimetic T423E mutation) in A549 cells induced ~10-fold higher viral titers compared to those observed after transfection of the control vector or of a plasmid expressing a kinase-dead (K299R) mutant PAK1. Furthermore, PAK1-specific siRNA reduced the virus yield by 10–100-fold, and treatment with TAT-PAK18, a cell-permeable anti-PAK1 peptide, suppresses both ERK1/2 phosphorylation and infectious virus production, as does U0126, a specific MEK/ERK inhibitor. These findings clearly indicate that like malaria and HIV infection, the influenza virus activates PAK1 in host cells during infection, which is essential for robust replication of the pathogen [49].

4.3.3 Human Papilloma Virus

As discussed in Chapter 2, HPV is a necessary cause of cervical cancer and a few other genital cancers [50]. Interestingly, like host cells infected by HIV and flu viruses, papillomas caused by HPV infection were recently shown to carry hyperactivated PAK1 and PAK2 through the ErbB1-RAC signal pathway [51]. In the case of HIV and flu viruses, their infection and replication require PAK1–3 in host cells. Thus, it is quite conceivable that HPV infection also requires PAK1/PAK2 in these genital cells. If that is the case, PAK blockers would be potentially far simpler therapeutics for these HPV-dependent cancers as well. In support of this notion, the estrogen-dependent growth of cervical cancer cells was shown to be strongly inhibited by the natural PAK1 inhibitor curcumin [52]. For the malignant growth of cervical cancer, HPV-encoding oncoproteins (E6 and E7) and the female steroid hormone estrogen are known to work synergistically [52], and as discussed in Chapter 2, estrogen receptor (ER) and PAK1 form a vicious oncogenic cycle.

Acknowledgments

The author is very grateful to Dr. Neal Alto for his kindly providing with Figures 4.1 and 4.2, Dr. Mike Rosen for Figure 4.3, and Drs. Jean-Philippe Semblat and Christian Doerig for their preliminary draft on a related subject, in particular malaria infection, which was eventually published in the PAK issue of *Cellular Logistics* [53].

References

[1] Singh AP, Buscaglia CA, Wang Q, Levay A, Nussenzweig DR, Walker JR, et al. Plasmodium circumsporozoite protein promotes the development of the liver stages of the parasite. Cell 2007;131:492–504.

[2] van de Sand C, Horstmann S, Schmidt A, Sturm A, Bolte S, Krueger A, et al. The liver stage of *Plasmodium berghei* inhibits host cell apoptosis. Mol Microbiol 2005;58:731–42.

[3] Hirokawa Y, Tikoo A, Huynh J, Utermark T, et al. A clue to the therapy of neurofibromatosis type 2: NF2/merlin is a PAK1 inhibitor. Cancer J 2004;10:20–6.

[4] Hirokawa Y, Arnold M, Nakajima H, Zalcberg J, et al. Signal therapy of breast cancer xenograft in mice by the HDAC inhibitor FK228 that blocks the activation of PAK1 and abrogates the tamoxifen resistance. Cancer Biol Ther 2005;4:956–60.

[5] Prudencio M, Rodrigues CD, Hannus M, Martin C, Real E, Goncalves LA, et al. Kinome-wide RNAi screen implicates at least 5 host hepatocyte kinases in *Plasmodium* sporozoite infection. PLoS Pathog 2008;4:e1000201.

[6] Haldar K, Mohandas N. Erythrocyte remodeling by malaria parasites. Curr Opin Hematol 2007;14:203–9.

[7] Buffet PA, Safeukui I, Deplaine G, Brousse V, Prendki V, Thellier M, et al. The pathogenesis of *Plasmodium falciparum* malaria in humans: insights from splenic physiology. Blood 2011;117:381–92.

[8] de Koning-Ward TF, Gilson PR, Boddey JA, Rug M, Smith BJ, Papenfuss AT, et al. A newly discovered protein export machine in malaria parasites. Nature 2009;459:945–9.

[9] Scherf A, Lopez-Rubio JJ, Riviere L. Antigenic variation in *Plasmodium falciparum*. Annu Rev Microbiol 2008;62:445–70.

[10] Koncarevic S, Rohrbach P, Deponte M, Krohne G, Prieto JH, Yates III J, et al. The malarial parasite *Plasmodium falciparum* imports the human protein peroxiredoxin 2 for peroxide detoxification. Proc Natl Acad Sci USA 2009;106:13323–13328.

[11] Harrison T, Samuel BU, Akompong T, Hamm H, Mohandas N, Lomasney JW, et al. Erythrocyte G protein-coupled receptor signaling in malarial infection. Science 2003;301:1734–6.

[12] Murphy SC, Harrison T, Hamm HE, Lomasney JW, Mohandas N, Haldar K. Erythrocyte G protein as a novel target for malarial chemotherapy. PLoS Med 2006;3:e528.

[13] Sicard A, Semblat JP, Doerig C, Hamelin R, Moniatte M, Dorin-Semblat D, et al. Activation of a PAK–MEK signalling pathway in malaria parasite-infected erythrocytes. Cell Microbiol 2011;13:836–45.

[14] Gardner MJ, Hall N, Fung E, White O, Berriman M, Hyman RW, et al. Genome sequence of the human malaria parasite *Plasmodium falciparum*. Nature 2002;419:498–511.

[15] Ward P, Equinet L, Packer J, Doerig C. Protein kinases of the human malaria parasite *Plasmodium falciparum*: the kinome of a divergent eukaryote. BMC Genom 2004;5:79.

[16] Anamika SN, Krupa A. A genomic perspective of protein kinases in *Plasmodium falciparum*. Proteins 2005;58:180–9.

[17] Deacon SW, Beeser A, Fukui JA, Rennefahrt UE, Myers C, Chernoff J, et al. An isoform selective, small-molecule inhibitor targets the autoregulatory mechanism of p21-activated kinase. Chem Biol 2008;15:322–31.

[18] Lu X, Wu X, Plemenitas A, Yu H, Sawai ET, Abo A, et al. CDC42 and Rac1 are implicated in the activation of the Nef-associated kinase and replication of HIV-1. Curr Biol 1996;6:1677–84.

[19] Arora VK, Molina RP, Foster JL, Blakemore JL, Chernoff J, Fredericksen BL, et al. Lentivirus Nef specifically activates Pak2. J Virol 2000;74:11081–11087.

[20] Sibony M, Jones NL. Recent advances in *Helicobacter pylori* pathogenesis. Curr Opin Gastroenterol 2012;28:30–5.

[21] Kusters JG, van Vliet AH, Kuipers EJ. Pathogenesis of *Helicobacter pylori* infection. Clin Microbiol Rev 2006;19:449–90.

[22] Liu Z, Xu X, Chen L, Li W, Sun Y, Zeng J, et al. *Helicobacter pylori* CagA inhibits the expression of Runx3 via Src/MEK/ERK and p38 MAPK pathways in gastric epithelial cell. J Cell Biochem 2012;113:1080–6.

[23] Baek HY, Lim JW, Kim H. Interaction between the *Helicobacter pylori* CagA and alpha-Pix in gastric epithelial AGS cells. Ann NY Acad Sci 2007;1096:18–23.

[24] Cunningham AL, Donaghy H, Harman AN, Kim M, Turville SG. Manipulation of dendritic cell function by viruses. Curr Opin Microbiol 2010;13:524–9.

[25] Segal ED, Cha J, Lo J, Falkow S, Tompkins LS. Altered states: involvement of phosphorylated CagA in the induction of host cellular growth changes by *Helicobacter pylori*. Proc Natl Acad Sci USA 1999;96:14559–14564.

[26] Churin Y, Kardalinou E, Meyer TF, Naumann M. Pathogenicity island-dependent activation of Rho GTPases Rac1 and Cdc42 in *Helicobacter pylori* infection. Mol Microbiol 2001;40:815–23.

[27] Lim JW, Kim KH, Kim H. alphaPix interacts with *Helicobacter pylori* CagA to induce IL-8 expression in gastric epithelial cells. Scand J Gastroenterol 2009;44:1166–72.

[28] Basak C, Pathak SK, Bhattacharyya A, Mandal D, Pathak S, Kundu M. NF-kappaB- and C/EBPbeta-driven interleukin-1beta gene expression and PAK1-mediated caspase-1 activation play essential roles in interleukin-1beta release from *Helicobacter pylori* lipopolysaccharide-stimulated macrophages. J Biol Chem 2005;280:4279–88.

[29] Selyunin A, Alto N. Activation of PAK by a bacterial type III effector EspG reveals alternative mechanisms of GTPase pathway regulation. Small GTPases 2011;2:217–21.

[30] Selyunin AS, Sutton SE, Weigele BA, Reddick LE, Orchard RC, Bresson SM, et al. The assembly of a GTPase-kinase signalling complex by a bacterial catalytic scaffold. Nature 2011;469:107–11.

[31] Glotfelty LG, Hecht GA. Enteropathogenic *E. coli* effectors EspG1/G2 disrupt tight junctions: new roles and mechanisms. Ann NY Acad Sci 2012;1258:149–58.

[32] Cheng HC, Skehan BM, Campellone KG, Leong JM, Rosen MK. Structural mechanism of WASP activation by the enterohaemorrhagic *E. coli* effector EspF(U). Nature 2008;454:1009–13.

[33] Cory GO, Cramer R, Blanchoin L, Ridley AJ. Phosphorylation of the WASP-VCA domain increases its affinity for the Arp2/3 complex and enhances actin polymerization by WASP. Mol Cell 2003;11:1229–39.

[34] Foster JL, Garcia JV. HIV-1 Nef: at the crossroads. Retrovirology 2008;5:84.

[35] Sawai ET, Baur A, Struble H, Peterlin BM, Levy JA, Cheng-Mayer C. Human immunodeficiency virus type 1 Nef associates with a cellular serine kinase in T lymphocytes. Proc Natl Acad Sci USA 1994;91:1539–43.

[36] Nunn MF, Marsh JW. Human immunodeficiency virus type 1 Nef associates with a member of the p21-activated kinase family. J Virol 1996;70:6157–61.

[37] Sawai ET, Khan IH, Montbriand PM, Peterlin BM, Cheng-Mayer C, Luciw PA. Activation of PAK by HIV and SIV Nef: importance for AIDS in rhesus macaques. Curr Biol 1996;6:1519–27.

[38] Renkema GH, Manninen A, Mann DA, Harris M, Saksela K. Identification of the Nef-associated kinase as p21-activated kinase 2. Curr Biol 1999;9:1407–10.

[39] Wolf D, Witte V, Laffert B, Blume K, Stromer E, Trapp S, et al. HIV-1 Nef associated PAK and PI3-kinases stimulate Akt-independent Bad-phosphorylation to induce anti-apoptotic signals. Nat Med 2001;7:1217–24.

[40] Krautkramer E, Giese SI, Gasteier JE, Muranyi W, Fackler OT. Human immunodeficiency virus type 1 Nef activates p21-activated kinase via recruitment into lipid rafts. J Virol 2004;78:4085–97.

[41] Hekman M, Albert S, Galmiche A, Rennefahrt UE, Fueller J, Fischer A, et al. Reversible membrane interaction of BAD requires two C-terminal lipid binding domains in conjunction with 14-3-3 protein binding. J Biol Chem 2006;281:17321–17336.

[42] Ayllon V, Fleischer A, Cayla X, Garcia A, Rebollo A. Segregation of Bad from lipid rafts is implicated in the induction of apoptosis. J Immunol 2002;168:3387–93.

[43] Polzien L, Baljuls A, Roth HM, Kuper J, Benz R, Schweimer K, et al. Pore-forming activity of BAD is regulated by specific phosphorylation and structural transitions of the C-terminal part. Biochim Biophys Acta 2011;1810:162–9.

[44] Rapp UR, Fischer A, Rennefahrt UE, Hekman M, Albert S. BAD association with membranes is regulated by Raf kinases and association with 14-3-3 proteins. Adv Enzyme Regul 2007;47:281–5.

[45] Nguyen DG, Wolff KC, Yin H, Caldwell JS, Kuhen KL. "UnPAKing" human immunodeficiency virus (HIV) replication: using small interfering RNA screening to identify novel cofactors and elucidate the role of group I PAKs in HIV infection. J Virol 2006;80:130–7.

[46] Xu JW, Ikeda K, Kobayakawa A, Ikami T, Kayano Y, Mitani T, et al. Downregulation of Rac1 activation by caffeic acid in aortic smooth muscle cells. Life Sci 2005;76:2861–72.

[47] Demestre M, Messerli SM, Celli N, Shahhossini M, Kluwe L, Mautner V, et al. CAPE (caffeic acid phenethyl ester)-based propolis extract (Bio 30) suppresses the growth of human neurofibromatosis (NF) tumor xenografts in mice. Phytother Res 2009;23:226–30.

[48] Messerli SM, Ahn MR, Kunimasa K, Yanagihara M, Tatefuji T, Hashimoto K, et al. Artepillin C (ARC) in Brazilian green propolis selectively blocks oncogenic PAK1 signaling and suppresses the growth of NF tumors in mice. Phytother Res 2009;23:423–7.

[49] Pascua PN, Lee JH, Song MS, Park SJ, Baek YH, Ann BH, et al. Role of the p21-activated kinases (PAKs) in influenza A virus replication. Biochem Biophys Res Commun 2011;414:569–74.

[50] zur Hausen H. Papillomaviruses in anogenital cancer as a model to understand the role of viruses in human cancers. Cancer Res 1989;49:4677–81.

[51] Wu R, Abramson A, Symons M, Steinberg B. Pak1 and Pak2 are activated in recurrent respiratory papillomas, contributing to one pathway of Rac1-mediated COX-2 expression. Int J Cancer 2010;127:2230–7.

[52] Singh M, Singh N. Curcumin counteracts the proliferative effect of estradiol and induces apoptosis in cervical cancer cells. Mol Cell Biochem 2011;347:1–11.

[53] Semblat JP, Doerig C. PAK in pathogen–host interactions. Cell Logist 2012;2:126–31.

5 PAK1 in Brain Diseases or Disorders

Hiroshi Maruta[1], and Shanta M. Messerli[2]

[1]NF/TSC Cure Organisation, Melbourne, Australia, [2]Marine Biological Laboratory, Woods Hole, MA, USA

Abbreviations

AD	Alzheimer's disease
ARC	artepillin C
BBB	blood–brain barrier
CA	caffeic acid
CAPE	caffeic acid phenethyl ester
CBZ	carbamazepine
CTCL	cutaneous T-cell lymphoma
DA	dopamine
DISC1	disrupted-in-schizophrenia1
DN	dominant negative
FST	forced swimming test
FXS	Fragile X syndrome
GPE	green propolis extract
HD	Huntington's disease
HDAC	histone deacetylase
KO	knockout
LD	learning deficit
MPNST	malignant peripheral nerve sheath tumor
NF	neurofibromatosis
NMDA	glutamate N-methyl-D-aspartic acid
RB	retinoblastoma
SEGAs	subependymal giant cell astrocytomas
SMEI	severe myoclonic epilepsy of infancy
TSC	tuberous sclerosis
UBE3A	ubiquitin protein ligase E3A
VPA	valproic acid

5.1 Introduction

Both class 1 PAKs (PAK 1–3) and class 2 PAKs (PAK 4–6) are highly expressed in mammalian brain, and there is no doubt about the anticipated physiological role of

PAKs, RAC/CDC42 (p21)-activated Kinases. DOI: http://dx.doi.org/10.1016/B978-0-12-407198-8.00005-9

each of these PAKs in a variety of brain functions. However, once the highly sophis-ticated brain network development is completed during embryogenesis, PAK1, at least, appears to be no longer necessary for the postnatal brain function of either the mouse or the tiny nematode *Caenorhabditis elegans* because PAK1-deficient mice and nematodes are quite healthy once born [1,2], although the litter size (number of eggs laid) is apparently reduced [3]. In the case of *C. elegans*, PAK1 and PAK2 (called MAX 2) are redundant functionally in part, and in the presence of PAK2, a PAK1-deficient strain (RB689) of this nematode is normal (and even more thermore-sistant), although double knockout (KO) of both PAK1 and PAK2 genes would make the nematode embryonically lethal or cause a severe neuronal deficit [2,3].

More importantly, a series of genetic and biochemical studies during the last two decades clearly demonstrated that abnormal activation of PAK1 in postnatal brain is predisposed to cause a variety of diseases or disorders such as malignant tumor (glioma) development, neurofibromatosis (NF), tuberous sclerosis (TSC), epilepsy, mental retardation/learning deficit (LD), depression and schizophrenia, in addi-tion to Alzheimer's disease (AD) and Huntington's disease (HD). Thus, too much kinase activity of PAK1 in brain is clearly pathogenic: it jeopardizes our quality of life and eventually could shorten our lifespan. In other words, a series of anti-PAK1 compounds or natural products such as propolis would be clearly beneficial for maintaining a variety of our neuronal functions and in improving our overall quality of life.

5.2   Tumorigenesis: NF, TSC, Retinoblastoma, and Gliomas

5.2.1   Neurofibromatosis

As discussed briefly in Chapters 2 and 3, dysfunction (loss-of-function) mutations of either of two specific tumor suppressor genes, *NF*1 or *NF*2, cause a variety of tumor-associated neuronal disorders, called NF type 1 or type 2, respectively [4,5]. Since the *NF*1 gene encodes an attenuator of both RAS and RAC of 2818 amino acids, dysfunction of this gene causes an abnormal activation of both RAS and RAC, and eventually leads to the pathogenic activation of PAK1 that causes *NF*1-deficient tumors such as malignant peripheral nerve sheath tumor (MPNST) and a few other types of benign tumors on skin, in brain, or along the spine, often leading to deafness, facial distortion, hunchback, or LDs such as low IQ [4,6]. On the other hand, the *NF*2 gene encodes a direct PAK1 inhibitor of 589 amino acids called mer-lin [5], and its dysfunction causes two types of brain tumors called meningioma and schwannoma, leading to blindness, deafness, and several other disabilities includ-ing total paralysis in the end (mainly due to repeated brain/spine surgeries). Both NF1 and NF2 are rare genetic diseases, developing in 1 in 3000 and 1 in 30,000 cases, respectively. Although the first NF patient was found in 1882 by Friedrich von Recklinghausen, the number of NF patients is so small in comparison to the num-ber of cancer patients that NF research has remained far behind rapidly advancing (and well–funded) cancer research, mainly because of a severe shortage of both NF scientists and NF research funds, and therefore no FDA-approved effective NF therapeutic is available on the market as yet.

However, shortly after the *NF1* and *NF2* genes were cloned in the early 1990s, we found that overexpression of either the *NF1* gene or the *NF2* gene can reverse RAS-induced malignant transformation [7,8], strongly suggesting that these rare NF tumors also share a common oncogenic signal pathway with RAS cancers such as pancreatic and colon cancers, in which Ki-RAS is oncogenically mutated. Thus, in principle, both RAS cancers and NF tumors could be potentially cured by means of a gene therapy technology using *NF1* or *NF2* genes, perhaps in the far future. Moreover, during 1997–1998, Jeff Field's group in Philadelphia found the very first clue as to a common oncogenic signal transducer that causes these formidable tumors. The growth of both RAS cancer and *NF1*-deficient tumors (MPNST) in mice could be suppressed by overexpression of a dominant negative (DN), kinase-dead mutant of PAK1 [9,10]. In other words, the growth of both RAS cancers and *NF1*-deficient tumors requires PAK1. However, until 2003, no human *NF2*-deficient tumor cell line that could grow in nude mice was available, and therefore nobody was able to test whether the growth of *NF2*-deficient tumors also requires PAK1.

Meanwhile, around mid-2002, a self-educated mother of an 8-year-old Sydney schoolboy with NF2 who suffered from meningioma, which causes loss of eyesight, urgently asked us to develop an effective NF therapeutic, as we had already identified a few potential RAS cancer therapeutics, such as CEP-1347 and a combination of Tyr kinase inhibitors (PP1 and AG 879), which block PAK1 selectively. The very first task of our team was to test if the *NF2* gene product merlin inhibits PAK1 directly or not. For by then we had biochemical evidence that merlin somehow blocks another kinase called JNK, downstream of PAK1, just like CEP-1347. On the analogy of CEP-1347, we then confirmed the anti-PAK1 activity of merlin both *in vitro* and *in vivo* [5]. Using CEP-1347 and a few other anti-PAK1 compounds, we found that the growth of *NF2*-deficient tumors such as mesothelioma and schwannoma also requires PAK1, as does the growth of RAS cancers and *NF1*-deficient MPNST.

Our next task was to test if the growth of *NF1*-deficient tumors (MPNST) is suppressed by FK228, the most potent histone deacetylase (HDAC) inhibitor, which we found eventually blocks PAK1 [11], and was in clinical trials (phase 2) for another rare cancer called cutaneous T-cell lymphoma (CTCL). Yes, FK228 (2.5 mg/kg, i.p., twice a week) completely shrank MPNST in nude mice without any side effect [12]. However, it was soon revealed that FK228 does not pass the blood–brain barrier (BBB), meaning that this potent drug is useless for rescuing patients suffering from brain tumors such as *NF2*-deficient meningioma and schwannoma as well as malignant gliomas, even if it comes on the market in the future. One can easily imagine how the mother of this NF2-suffering young boy and many other NF2 families were disappointed to hear this bittersweet news.

So we decided to take an entirely unorthodox approach to explore effective NF2-therapeutics: we would try to identify anti-PAK1 drugs among natural products inexpensively available on the market, a strategy we would not be allowed to pursue at our the so-called "ivory tower," where we had conducted only basic cancer research for almost two decades. For this new mission of ours, in early 2006, one of us

(Dr. Maruta) moved to Hamburg University Hospital (UKE), which is a Mecca for NF clinical research and therapy in Europe. There we started to work on a bee product called propolis, a very old traditional medicine for the treatment of infectious disease and inflammation, used since the ancient Egyptian era. Fortunately, in Germany, propolis is officially registered as a clinical drug, although it is treated as just an alternative medicine in other developed countries such as the United States, Japan, Oceania, and the rest of Europe. Around 1988, a group at Columbia University found that a propolis extract contains an anticancer ingredient called caffeic acid phenethyl ester (CAPE), which can be chemically or enzymatically synthesized from caffeic acid (CA) and phenylalcohol (PA) (Figure 5.1) [13]. Since then many terminal cancer patients, who had lost their faith in highly toxic chemotherapies (DNA/RNA/ microtubule poisons), began to take propolis as an alternative medicine with great success. However, until recently nobody knew how propolis selectively kills cancers without causing any side effects. In 2005, a Japanese group reported for the first time that CA downregulates the GTPase RAC, a direct activator of PAK1 [14], strongly suggesting that its derivative CAPE in propolis could also block PAK1 selectively.

However, CAPE content in propolis varies greatly from one sample to another, depending on the climate, area, and season when bees harvest it. For instance, Brazilian green or red propolis samples contain no CAPE but are still anticancerous because of other anticancer ingredients such as artepillin C (ARC) and triterpenes [15,16]. According to a 2002 poster from a group at Sydney University, New Zealand (NZ), propolis contains CAPE at the highest level (around 6–7% in dry extract) found in propolis samples from around the world. So we decided to work on an NZ propolis extract called Bio 30, produced by Manuka Health in Auckland, to test its therapeutic effect on both human *NF*1-deficient malignant tumors (MPNST) and *NF*2-deficient tumors (schwannoma) grafted in nude mice [17], in addition to

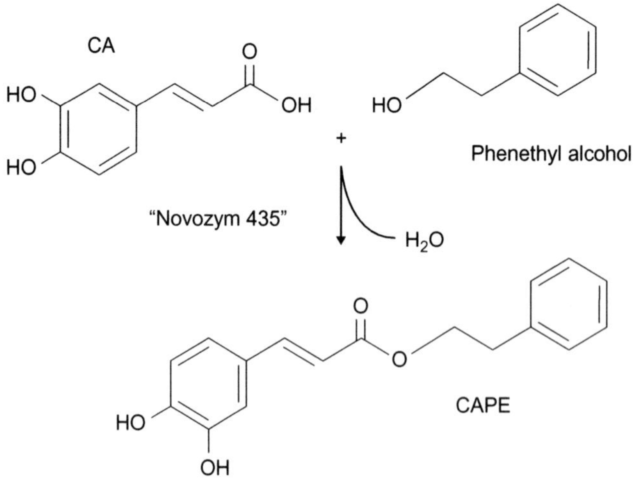

Figure 5.1 Enzymatic synthesis of CAPE.

human pancreatic and breast cancers and glioma [18]. Bio 30 (100 mg/kg, i.p., twice a week) completely blocked the growth of all these tumors, and in the case of *NF2*-deficient schwannoma, in a month, this tumor disappeared almost completely. In addition, we found that both Brazilian green propolis extract (GPE) and its major anticancer ingredient ARC selectively block PAK1 without affecting another oncogenic kinase, AKT, and strongly suppress the growth of *NF2*-deficient schwannoma grafted in mice, although unlike Bio 30, GPE (500 mg/kg, i.p., twice a week) could not shrink the tumors [19].

It should be worth noting that in GPE, ARC is the sole anticancer ingredient, whereas in Bio 30, several anticancer ingredients including CAPE, CA, apigenin, pinocembrin, galangin, and chrysin are present (Table 5.1), and these polyphenols work synergistically with CAPE and boost the antitumor activity of CAPE by around 600 times *in vitro* [17]. This synergy is most likely one of the major reasons why Bio 30 is apparently far more potent than the ARC-based GPE against *NF2*-deficient schwannoma, at least. Another reason is that the IC_{50} of ARC for this schwannoma cell line is around 210 μM, which is several times higher than the IC_{50} (around 25 μM) for the MPNST cell line [19]. However, the opposite is also true for CAPE-based Bio 30. The IC_{50} of Bio 30 of the MPNST cell line is around 8 μg/ml, several times higher than that of the schwannoma cell line(1.5 μg/ml) [17]. In other words, it is conceivable that the ARC-based GPE could be as effective as Bio 30 for MPNST therapy, although we have never directly compared the effect of GPE and Bio 30 on MPNST *in vivo*.

As shown in Table 5.1, the CAPE content in Bio 30 is apparently far lower than 6–7% (60–70 mg/g), the quantity that was reported in the 2002 poster. Nobody knows the precise reason why, but a propolis expert suspects that since the CAPE content shown in 2002 poster is based on thin layer chromatography analysis, this rather low-technology method might not be sensitive enough to separate CAPE from the more abundant pinobanksin 3-acetate, which could be separable using more sophisticated reverse liquid chromatography. Another possibility could be a rapid hydrolysis of CAPE into CA and PA during the first step of propolis preparation for the removal of the water-insoluble wax in a water bath. To further test the therapeutic effect of a higher CAPE-content propolis sample, we created a legendary CAPE 60 (containing 60 mg of CAPE per gram) by adding synthetic CAPE to Bio 30. This CAPE 60 turned out to be far more effective than Bio 30 alone in suppressing both

Table 5.1 Polyphenol Content (mg/g): Bio 30 and Chinese Red Propolis Extract (RPE)

Polyphenols	Bio 30 (17)	Chinese RPE (18)
CAPE	12	17
CA	12	13
Pinocembrin	110	84
Galangin	60	37
Chrysin	30	50
Apigenin	12	Not determined

metastasis and growth of MPNST in mice [17], clearly confirming that the therapeutic effect of CAPE-based propolis depends on the CAPE content.

Based on the encouraging outcome of these animal tests, around mid-2007, we started conducting a worldwide trial of Bio 30 (alcohol-free liquid) containing 250 mg of extract per milliliter, mainly for NF patients, to see if Bio 30 also works on human patients. The outcome of this human trial, conducted during the last 5 years in over 200 NF patients, is summarized below.

The effective minimum daily dose of Bio 30 (25 mg/kg = 1 ml/10 kg) has stopped the growth of tumors in most NF1 and NF2 patients, as well as in glioma and pancreatic cancer patients, without any side effect [18]. Furthermore, in three NF1 (dermal neurofibroma) cases, tumors completely disappeared in a month. Also, in at least three cases of NF2 (both schwannoma and meningioma) and one glioma case, these brain tumors shrank by more than 50% in two to three years. In two pancreatic cancer cases, both early and metastasized terminal cancers completely disappeared in one year [18]. In another NF2 patient's case, his lost hearing and sense of balance came back in several months with Bio 30 treatment, and an MPNST patient treated with Bio 30 has had no relapse of MPNST in the last three years (Maruta H., unpublished observation). These preliminary clinical trial data strongly suggest that Bio 30 is effective in suppressing the growth of both NF1 and NF2 tumors as well as glioma and pancreatic cancers, against which the conventional anticancer drugs are basically useless. However, it should not be surprising that unlike the Bio 30 sensitivity of NF/cancer-carrying cloned nude mice, which are genetically identical, the sensitivity in human NF/cancer patients varies vastly from one person to another, depending on the difference in their symptoms *per se*, as well as on their genetic background. For instance, cure of an early pancreatic cancer requires only 25 mg/kg of Bio 30 daily for a year, but cure of a terminal metastasized pancreatic cancer requires 200 mg/kg of Bio 30 for a year. The great advantage to Bio 30 is that even such a massive daily dose causes no side effects. Likewise, shrinking a few wart-like dermal NF1 tumors with 25 mg/kg daily takes only a few weeks, while shrinking either large plexiform NF1 tumors or NF2-associated brain tumors with the same minimum daily dose takes two to three years.

Although we have not done any systematic clinical trials of GPE for NF patients, at least in two NF1 (dermal neurofibroma) cases, GPE (8 mg/kg, daily) was sufficient to shrink these skin tumors completely in a month or so, suggesting that GPE is also useful for NF (including MPNST) patients. For the last decade, GPE has been used in human trials for a variety of PAK1-dependent cancers such as pancreatic, colon, gastric, breast, prostate, lung, liver, ovarian, cervical, and thyroid as well as glioma, melanoma, and multiple myeloma, which represent more than 70% of cancers, and in many cases, GPE has been proven to be effective in either stopping their growth or shrinking these solid cancers. Thus, we trust that GPE would be useful for both NF1 and NF2 patients as well. The only minor problem with GPE is that it costs several times as much as Bio 30, which costs only a dollar daily for a lifelong treatment of NF. On the other hand, like any other CAPE-based propolis, Bio 30 causes an allergic skin reaction in 1% of the population, while GPE causes no allergic reaction.

It should be worth emphasizing that our human trials are not the standard double-blind clinical tests that are usually conducted for the FDA approval of drug marketing by pharmaceutical companies. Our specific humane reason (ethical principle) is as follows. The patients in our trials, who are suffering from NF or other PAK1-dependent solid tumors, simply want to stop the growth of their tumors or get rid of them once and for all, if possible without any side effects, and do not wish to take any placebo, which has no therapeutic effect at all. If we had divided these patients into two groups, and given a placebo to one group and Bio 30 only to the other, we would have betrayed the faith of the entire first group, and such an inhumane treatment would not be justified among the NF- or cancer-suffering community. Since we have already conducted a strict double-blind test of Bio 30 on mice, which cannot distinguish between a placebo and Bio 30, it is absolutely clear that the observed positive therapeutic effect is based on Bio 30 *per se* and not the so-called placebo (psychological) effect. Unlike nude mice that have been cloned and bred only for testing the anticancer property of each new preclinical drug, human patients should never be used as redundant guinea pigs.

Since the most potent PAK1 blocker, FK228, was recently marketed by Celgene for the rare lymphomas CTCL and peripheral T-cell lymphoma under the brand name Istodax, we are planning to start the humane clinical trials of Istodax versus propolis (GPE or Bio 30) for NF1 (in particular MPNST) patients soon, in order to extend the FDA approval of Istodax to NF1 therapy in the future.

5.2.2 Tuberous Sclerosis

TSC is another rare tumor-associated genetic disease or disorder that occurs not only in the brain, often causing severe epilepsy, but also in the kidney and heart. The currently available therapeutic approaches are surgical removal of TSC tumors and antiepileptic (anticonvulsant) drugs such as carbamazepine (CBZ), oxcarbazepine, and valproic acid (VPA), which only treat the epilepsy *per se*. Like NF, TSC develops in a very early stage of our life (around six months after birth), and its conditions progressively worsen with age if not properly treated. Unfortunately, however, no FDA-approved systemic therapeutic that could cure TSC is available on the market as yet.

There are two distinct tumor suppressor genes, *TSC1* and *TSC2*, whose dysfunction (loss-of-function mutation) causes TSC in around 1 in 5000 humans. *TSC1* and *TSC2* gene products called harmartin and tuberin form a TSC complex [20]. This complex is a GAP (attenuator) of another oncogenic GTPase called Rheb [21]. As shown in Figure 5.2, once Rheb is abnormally activated, it activates another oncogenic kinase called target of rapamycin (TOR) [22]. The antibiotic rapamycin is known to inhibit TOR [21]. Thus, in principle, rapamycin or its derivatives could serve as effective therapeutics for TSC, as proven in mouse or Eker rat models [23,24]. Unfortunately, however, rapamycin and its derivatives such as Afinitor (RAD001) are rather toxic, and cause a few side effects, such as immunosuppression and hypertension, in human beings. Furthermore, so far Afinitor (from Novartis) is the first and sole TOR inhibitor. It became available on the market in late 2010, and

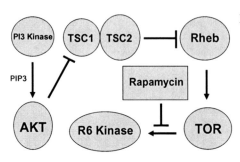

Figure 5.2 TSC complex blocks TOR.

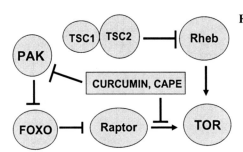

Figure 5.3 Blocking raptor-TOR signaling.

is the first medication for subependymal giant cell astrocytomas (SEGAs) [25]. As it is extremely expensive, mainly because of the monopoly status, it would bankrupt TSC patients and their families because the treatment of TSC should be lifelong. Notably, SEGAs occur only in 20% of TSC patients.

Is there any potentially effective alternative medicine for TSC? Yes, anti-PAK1 drugs would be an alternative, but the link of PAK1 to TSC is a rather long and complex story. The full activation of TOR needs not only Rheb but also another protein called raptor [26]. This essential TOR-raptor interaction is blocked by curcumin [27], one of the direct PAK1 inhibitors, clearly indicating that the TOR-raptor interaction requires PAK1 (Figure 5.3). Curcumin, in fact, suppresses epilepsy in a mouse model [28]. Furthermore, Bio 30, an NZ propolis extract that is rich in the PAK1 blocker CAPE, also stopped epilepsy in a few TSC patients (children) in our trials [18]. Also it should be pointed out that one of the widely used antiepileptic drugs, VPA, is an HDAC inhibitor that eventually blocks PAK1 [29] and extends the lifespan of *C. elegans* by activating FOXO [30].

Interestingly, 10-hydroxy 2-decenoid acid (10HDA), the major fatty acid in royal jelly, is an HDAC inhibitor [31], therefore eventually blocking PAK1 as FK228, and in fact extends the lifespan of *C. elegans* [32]. Royal jelly is a literally royal food; it is prepared by worker bees for the queen bee and is thought to be the major contributor to her much longer lifespan (three to four years); worker bees only live several months. Although royal jelly is far more expensive than propolis, we trust that royal jelly as well as all the other PAK1 blockers discussed in Chapter 3 or 7 would be potentially useful for lifelong TSC treatment.

5.2.3 Retinoblastoma

The *RB* gene is the first cloned tumor suppressor gene whose dysfunction (loss-of-function mutation) causes an eye cancer called retinoblastoma (RB) in young children [33]. The *RB* gene product is a transcription factor that normally blocks cell cycling (shift from G1 phase to S phase) by binding the C-terminus of another transcription factor called E2F-1, essential for the initiation of DNA replication [34]. However, when the RB protein is phosphorylated by cyclin-dependent kinases (CDKs), its function (binding to both the specific DNA sequence and E2F-1) is lost, and DNA replication is initiated [35]. Many oncogenic gene products such as RAS and PI-3 kinase activate CDKs by upregulating cyclin D1, which in turn activates CDKs [36]. Since PAK1 is activated by these oncogenic proteins and is essential for upregulation of cyclin D1 [37], PAK1 blockers such as CEP-1347, FK228, and propolis could restore the G1-arresting function of RB protein by downregulating cyclin D1 [37]. In fact, it was reported that butyric acid, an HDAC inhibitor, blocks the growth of malignant RB cells, at least *in vitro* [38]. As described in Chapter 3, HDAC inhibitors such as FK228 eventually block PAK1 [11]. Thus, in principle, all PAK1 blockers listed in Chapters 3 and 7 would be potentially useful for the RB treatment in the future. However, no FDA-approved effective RB therapeutic is available on the market as yet.

5.2.4 Glioma

A malignant brain tumor called glioma is derived from glial cells and develops in brain or spine. Several acquired (not inherited) genetic mutations have been found in gliomas. Among these mutations, either dysfunction (loss-of-function mutation) or DN mutation of the tumor suppressor p53 occurs, often at an early stage [39]. p53 is the guardian of the genome, which, during DNA and cell duplication, makes sure that the DNA is copied correctly and destroys the cell (through apoptosis) if the DNA is mutated and cannot be fixed. Thus, when p53 itself is mutated, mutations of other genes can survive. It should be worth noting that p53 mutation alone does not cause malignant transformation, and it has been well established that mutation of a few additional genes such as the oncogenic RAS mutation is required for malignancy to be induced by p53 mutation [40]. In other words, it has been suggested that either AKT or PAK1, two major oncogenic kinases downstream of PI-3 kinase, which is directly activated by RAS, is essential for the growth of gliomas.

In 1997, when another tumor suppressor called PTEN was first cloned, around 30% of gliomas were found to contain the dysfunction (loss-of-function mutation) of this unique PIP3 phosphatase, serving as an antagonist of PI-3 kinase, while almost all prostate cancers contain the dysfunction of PTEN [41]. The dysfunction of PTEN leads to abnormal activation of both PAK1 and AKT. The amplification of a Tyr kinase called EGFR/ErbB1, a cell surface receptor for the growth factor EGF, is often found in gliomas, and is responsible for their malignancy [42]. ErbB1 activates RAS, which in turn activates PI-3 kinase, eventually activating both PAK1 and AKT.

FK228, a potent HDAC inhibitor that eventually blocks PAK1, was shown to suppress the growth of human gliomas (e.g., U87-MG) in which ErbB1 is amplified, grafted in nude mice [43], strongly suggesting that the growth of gliomas requires PAK1. As discussed in Chapter 3, FK228 is impermeable through the BBB. Thus, clinically FK228 would be almost useless for therapy of brain tumors in general. However, to our surprise, it was found recently that glioma disrupts the BBB [44], suggesting the possibility that FK228 may be potentially useful for glioma therapy.

CAPE is the first among the BBB-permeable PAK1 blockers to be shown to block the growth of human glioma (C6) grafted in nude mice [45]. CAPE (1–10 mg/kg, i.p.) significantly reduced the size of this glioma. Thus, although CAPE alone cannot be used clinically, mainly because of its poor bioavailability, CAPE-rich propolis extracts such as Bio 30, currently available on the market, would be potentially useful for glioma therapy.

Furthermore, curcumin (100 mg/kg), a direct PAK1 inhibitor, was also shown to suppress the growth of human gliomas (U87-MG etc) grafted in mice [46], confirming that gliomas require PAK1 for their growth. However, as described in Chapter 3, just like CAPE, curcumin alone cannot be clinically used because of its poor bioavailability. Finally, we were able to demonstrate that Bio 30 (100 mg/kg, i.p. twice a week) indeed blocks the growth of human glioma almost completely (U87-MG) in a mouse xenograft model [18]. Thus, in our Bio 30 trials, we currently use Bio 30 for the treatment of several glioma patients, and at least in one case, the glioma shrank significantly with Bio 30, and the condition of this patient clearly improved [18].

5.3   Epilepsy

Around 50 million people worldwide suffer from epilepsy. Seizure or epilepsy is associated with a variety of brain tumors such as TSC [47]. Sixty to ninety percent of TSC patients develop epilepsy. Several other genetic brain diseases such as Dravet's syndrome also cause epilepsy. Dravet's syndrome, also known as severe myoclonic epilepsy of infancy (SMEI), is a neurodevelopmental disorder that begins in infancy and is characterized by severe epilepsy that does not respond well to any treatment. This syndrome was first described in 1978 by Charlotte Dravet, French psychiatrist and epileptologist, working then at the Centre Saint Paul at the University of Marseille. This rare disorder occurs in 1 in 20,000 births. Seventy to eighty percent of SMEI patients carry a loss-of-function mutation of the sodium channel $\alpha 1$ subunit gene. Epilepsy is usually controlled, but not cured, by treatment with antiepileptic (anticonvulsant) drugs such as CBZ and VPA. Unfortunately, however, SMEI is quite resistant to both CBZ and VPA. Thus, a combination of clobazam and stiripentol, which increases GABA transmission, is currently used for the treatment of SMEI instead. For details, see the following Web site: http://adreamforsarah.com/dravet_syndrome.php

CBZ was discovered in 1953 by Walter Schindler at Geigy AG (now called Novartis) in Switzerland [48]. He synthesized the drug in 1960 before its antiepileptic properties had been discovered. CBZ was first marketed in 1962 as a drug to treat trigeminal neuralgia (formerly known as tic douloureux). It has been used as an anticonvulsant in the United Kingdom since 1965 and was approved in the United States in 1974.

VPA was first synthesized in 1882 by B.S. Burton as an analog of valeric acid, found naturally in valerian. Interestingly, VPA is a *fatty acid* that is a clear liquid at room temperature. For many decades, it was used only in laboratories as a metabolically inert solvent for organic compounds. In 1962, however, the French researcher Pierre Eymard serendipitously discovered the anticonvulsant properties of VPA while using it as a vehicle for a number of other compounds that were being screened for antiseizure activity. He found that it prevents pentylenetetrazol-induced convulsions in rats [49]. It was approved as an antiepileptic drug in 1967 in France and has become the most widely prescribed antiepileptic drug worldwide.

The major common anticonvulsant mechanism of both CBZ and VPA still remains to be conclusively determined. However, it is now clear that like butyrate, both CBZ and VPA are among HDAC inhibitors [29,38,50], although their IC_{50} is rather high, around $2\,\mu M$ (compared with that of FK228, around $1\,nM$), and they eventually block the oncogenic kinase PAK1. Unfortunately, however, the potent FK228 cannot be used as an anticonvulsant drug because it does not pass the BBB. Thus, we have successfully treated TSC children whose epilepsy turns out to be quite resistant to conventional anticonvulsant drugs such as VPA and CBZ with Bio 30, a CAPE-based propolis which blocks PAK1 potently without any side effects [18].

Mutations in several distinct genes that code for protein subunits of voltage-gated and ligand-gated ion channels have been associated with forms of generalized epilepsy and infantile seizure syndromes. Several ligand-gated ion channels have been linked to some types of frontal and generalized epilepsy. It is speculated that one mechanism for some forms of inherited epilepsy is mutation of the genes that code for sodium channel proteins; these defective sodium channels stay open for too long, thus making the neurons hyperexcitable. Glutamate, an excitatory neurotransmitter, may be released from these neurons in large amounts, which—by binding with nearby glutamatergic neurons—triggers excessive calcium (Ca^{2+}) release in these postsynaptic cells. Such excessive calcium release can be neurotoxic to the affected cell. Interestingly, the oncogenic kinase PAK1 is required for such calcium mobilization [1]; this explains why PAK1 blockers such as VPA, CBZ, and propolis could effectively ease a variety of epileptic symptoms. The hippocampus, which contains a large volume of just such glutamatergic neurons as well as N-methyl-D-aspartic acid (NMDA) receptors, which are permeable to Ca^{2+} entry after binding of both glutamate and glycine, is especially vulnerable to epileptic seizure, subsequent spread of excitation, and possible neuronal death. Another possible mechanism involves mutations leading to ineffective action of GABA, the brain's most common inhibitory neurotransmitter. Epilepsy-related mutations in some nonion channel genes have also been identified. Among these genes are the *TSC1/TSC2* and *FXS* genes.

Furthermore, epilepsy occurs with increased frequency in people with an intellectual disability or LD compared to the rest of the population [51]. Among these LD syndromes are fragile X syndrome (FXS), Rett syndrome (RTT), and Angelman syndrome as well as TSC and SMEI.

These LD syndromes can eventually lead to alterations in the functioning of the glutamatergic and GABAergic neurotransmitter systems. The mechanisms involved include transcriptional regulation of the *RTT* gene and dysfunction (loss-of-function mutation) of the *FXS1*, *TSC1/TSC2*, and ubiquitin protein ligase (*UBE3A*) genes. Expression or functioning of receptor subunits, uptake sites, and enzymes involved in neurotransmitter metabolism are often affected by these changes and may lead to modifications in network excitability and neuronal plasticity, which may contribute to epileptogenesis and LD. As discussed below, the LD in NF1, TSC, and FXS as well as Down syndrome appears to be caused in part by hyperactivation of the onco-genic kinase PAK1 or TOR. However, the intriguing fact that no NF2 patients suffer from LD clearly indicates that the deregulation of PAK1 or TOR alone is not suffi-cient to cause LD, although it is an essential part of LD. In support of this notion, as discussed later, although Isaac Newton and John Nash suffered from schizophrenia, which also requires PAK1, neither of these mathematic geniuses had LD. Thus, it is quite conceivable that PAK1-deficient mice [1] would be resistant to a variety of PAK1-dependent brain diseases such as epileptogenesis, LD and schizophrenia, as well as all sorts of solid tumors.

5.4   Mental Retardation

Mental retardation or LD is associated with a variety of genetic diseases such as NF1, TSC, FXS, and Down syndrome (DS). In at least the first three diseases, abnor-mal activation of PAK1 is directly or indirectly involved [10,27,52]. For instance, in an NIH group's study performed two decades ago, around half of the NF1 children suffered from LD and had a significantly lower full-scale IQ score than the control (healthy) children [53]. In addition, most NF1 children suffer from visual spatial ori-entation deficit [53].

Trisomy 21 (Ts21) is among the most complex genetic conditions compatible with human survival past term. Ts21 results in Down syndrome (DS), which is the most common genetic cause for cognitive impairment, occurring at a frequency of around 1 in 700 live births [54]. Dosage imbalance of more than 300 genes [55] that are present on the extra copy of chromosome 21 results in a wide range of clini-cal features, including hypotonia, speech and language impairment, congenital heart defects, and craniofacial dysmorphology [56]. Achieving improvements in cognitive ability that would expand the potential of people with DS to live more independently has been the goal of decades of research in the field.

Recently, a Spanish group found that in a DS mouse model, the oncogenic PI-3 kinase–AKT–TOR signaling pathway is abnormally activated, and rapamycin, a TOR inhibitor, blocks the abnormal activation of these kinases [57], suggesting the possibility that either TOR inhibitors or PAK1 blockers, which eventually block TOR, could be potentially useful for DS therapy in the future.

In 2007, Susumu Tonegawa's group at MIT first found that the genetic disorders (abnormalities) such as mental retardation and autism associated with FXS require the kinase PAK1 [52]. In a mouse model where the *fragile X mental retardation* 1 (*FMR1*) gene is silenced (KO), FXS-associated abnormalities are rescued, at least partially, at both cellular and behavioral levels, by an inhibition of the kinase activ-ity of PAK1, which plays a critical role in actin polymerization and dendritic spine morphogenesis [58]. Greater spine density and elongated spines in the cortex, morphological synaptic abnormalities commonly observed in FXS, are at least partially restored by postnatal expression of a DN mutant of PAK1 in the forebrain. Likewise, the deficit in cortical long-term potentiation observed in *FMR1* KO mice is fully restored by this DN mutant of PAK1. Several behavioral abnormalities associated with *FMR1* KO mice, including those involving locomotor activity, stereotypy, anxiety, and trace fear conditioning are also partially or fully ameliorated by this PAK1 mutant. Finally, they demonstrated a direct interaction between PAK1 and the *FMR1* gene product *in vitro*, suggesting the possibility that like merlin (an *NF2* gene product), the *FMR1* gene product is a PAK1 blocker (and therefore a tumor suppressor). It was also reported by others that the *FMR1* gene product downregulates RAC, an activator of PAK1, leading to the inactivation of PAK1 [59]. These findings, taken together, strongly suggest that anti-PAK1 drugs will be among future FXS and autism therapeutics.

Shortly after this 2007 discovery, Susumu Tonegawa (the 1987 Nobel laureate) cofounded a venture biotech company called Afraxis in La Jolla/San Diego to develop a series of potent PAK1-specific inhibitors. In 2011, this company patented a few potent PAK1-specific inhibitors (IC_{50} around 10 nM) for the treatment of CNS disorders (http://www.sumobrain.com/patents/wipo/8-2-heterocycylpyrido23-dpy-rimidin-78h/WO2011156775.html).

These inhibitors could be taken orally, and pass BBB, according to the following Web site of Children's Tumor Foundation for NF people: http://www.ctf.org/pdf/ddi/2010-07-002.pdf.

Thus, in a distant future, these potent PAK1 inhibitors could be potentially useful not only for this type of autism but also for several other PAK1-dependent brain diseases such as NF, TSC, epilepsy, LD, depression, schizophrenia, AD, and HD, in addition to more than 70% of cancers (in particular solid tumors). Mysteriously, however, so far none of their specific chemical code names, structures, or further biological *in vitro* or *in vivo* data are available in any publication listed in PUBMED.

How does PAK1 control synapse formation? Three of seven recently identified genes mutated in nonsyndromic mental retardation are involved in the GTPase RAC. Two of the gene products, PIX (PAK interacting exchange factor) and PAK3, form a complex with a synaptic adaptor protein called GIT1 (G-protein-coupled receptor kinase-interacting protein 1). Using an RNA interference approach in 2005, Alan Horwitz's group at the University of Virginia found that GIT1 is critical for spine and synapse formation [58]. They also demonstrated that RAC is locally activated in dendritic spines using fluorescence resonance energy transfer. This local activation of RAC is regulated by PIX. PAK1 and PAK3 serve as downstream effectors of RAC in regulating spine and synapse formation. Active PAK promotes the formation of spines and dendritic protrusions, which correlates with an increase in the number of excitatory synapses. These effects depend on the kinase activity, phosphorylating the myosin II regulatory light chain. Thus, PAK stimulates not only smooth muscle contraction in stomach and blood vessels but also synapse function of brain through the myosin II ATPase. Activated myosin II causes an increase in dendritic spine and synapse formation, whereas inhibition of myosin II results in decreased spine and synapse formation. Finally, both activated PAK and myosin II can rescue the defects of GIT1 KO, suggesting that PAK and myosin II are downstream of GIT1 in regulating spine and synapse formation. These results point to a signaling complex, consisting of GIT1, PIX, RAC, and PAK, that plays an essential role in the regulation of dendritic spine and synapse formation and provides a potential mechanism by which PIX and PAK3 mutations affect cognitive functions in mental retardation.

5.5   Schizophrenia

As described in detail by Sylvia Nasar in her 1998 biography *A Beautiful Mind*, John Nash, a mathematical genius at Princeton University, suffered from schizophrenia for several decades of his career, and eventually recovered from it just before he shared the 1994 Nobel Prize in Economics [60]. According to Richard Westfall's 1993 book, *The Life of Isaac Newton*, this great British mathematical genius also suffered from schizophrenia at age 51, although he recovered from it later [61]. Eugen Bleuler (1857–1939), the Swiss psychiatrist, who coined the term *schizophrenia* in 1908, described a specific type of alteration of thinking, feeling, and relation to the external world [62]. The term refers to a splitting of psychic functions, a peculiar destruction of the inner cohesiveness of psychic personality.

A recent report from a Chinese group led by Hwai-Jong Cheng at the University of California at Davis (UCD) drew our attention to schizophrenia, a neurodevelopmental disorder with a genetic predisposition, for almost the first time in our lives. The reason is very simple: Using *C. elegans* as the tiniest (and simplest) model, they discovered a specific role of the GTPase RAC and its effector PAK1 in schizophrenia [63]. Schizophrenia is caused by dysfunction (loss-of-function mutation) of a gene called *DISC1* (disrupted-in-schizophrenia1), presumably among several other distinct genes [64]. Since this worm lacks any gene homologous to the mammalian *DISC1* gene, they generated a transgenic worm expressing a presumably DN mutant of DISC1 that is linked to GFP, and found that DISC1-GFP is localized in the growth cone and that this transgenic worm exhibits axon guidance defects. Furthermore, they found that the DISC1 mutation activates RAC–PAK1 signaling in this worm. In mammals, DISC1 interacts with TRIO, a RAC activator, that activates the kinase PAK1 [63]. In other words, DISC1 normally blocks TRIO–RAC–PAK1 signaling in mammals. Thus, like merlin (an *NF2* gene product), DISC1 is a tumor suppressor that blocks the oncogenic kinase PAK1. Thus, we presume that PAK1 is abnormally activated in the brain of schizophrenia patients with DISC1 mutation, and that in principle, anti-PAK1 drugs could suppress schizophrenia as well. In fact, according to a 2010 review by an Indian group, among PAK1 blockers, berberine, at least, appears to suppress schizophrenia as well as other PAK1-dependent neuronal disorders such as depression [65]. Thus, it would be worth testing the therapeutic effect on DISC1-induced schizophrenia of other natural PAK1 blockers such as propolis and curcumin.

In this context, it should be worth noting that PP1, an inhibitor of the SRC family Tyr kinases, was found in adult rats to suppress both hyperactivity and injury of retrosplenial cortical neurons produced by MK-801, a glutamate (NMDA) antagonist [66]. As discussed in Chapter 3, PP1 inactivates PAK1 by blocking the Tyr kinase FYN. Like other NMDA blockers such as ketamine, MK-801 causes a schizophrenia-like psychosis in humans [66]. The MK-801-induced hyperactivity and neuron injury are associated with induction of a heat-shock protein called HSP70. Interestingly, however, MK-801 induction of HSP70 is suppressed by the pretreatment of rats with either PP1 (0.1 mg/kg) or several atypical antipsychotics such as ziprasidone (32 mg/kg) that induce dopamine (DA) release. (Interestingly, Ziprasidone causes a spontaneous orgasm in a bipolar patient [67].) These observations hint at a close link between ecstasy (happy feeling) and suppression of PAK1.

5.6 Depression

We have been wondering for several years—since we started working with an old bee-made antibiotic called propolis—why or how bees have (or have acquired) a unique instinct to collect anti-PAK1 compounds for preparing propolis wherever they live, whether that is in the Far East, Oceania, Europe, or Brazil. Recently, we got a feeling that we have finally reached a very interesting conclusion, or at least taken the first step toward unraveling this centuries-old mystery.

A friend of ours working at the API Research Center in Japan recently informed us that CAPE, the major anticancer ingredient and anti-PAK1 compound in propolis, could work as an effective antidepressant, at least in an animal model (Ichihara, K. et al., personal communication). So we started looking through PUBMED for a few other herbal antidepressants, and found apigenin in propolis, curcumin in Indian curry, berberine in Chinese yellow root, and salidroside in *Rhodiola rosea* (golden root). Very interestingly, what is in common to these herbal antidepressants is their ability to block the oncogenic kinase PAK1 and activate the tumor-suppressing kinase AMPK.

According to a 2003 paper published by Keisuke Ohsawa's group in Sendai, apigenin (100 mg/kg) increases the pleasure hormone DA and improves impaired behaviors, such as the duration of immobility caused by a forced swimming test (FST) in depressed mice [68]. In other words, apigenin makes these depressed mice happy again by increasing DA production or secretion, just as alcoholic drinks such as beer and wine do for humans in whom the sensitivity (or number) of DA receptors is rather low. In 2005, a very similar antidepressant (DA boosting) effect of curcumin (10 mg/kg) was reported by a group at Peking University [69]. In 2007, a similar antidepressant effect of berberine (20 mg/kg) was found in mouse models by a Taiwanese group at China Medical University [70].

In 2008, a Swedish group found that the adaptogen *Rhodiola rosea* extract (20 mg/kg) showed a strong antidepressant effect on the duration of immobility of rats in a FST. Its major effective ingredient is salidroside, a hydroxyl-phenethyl glucoside [71]. Furthermore, in 2010, a Japanese group led by Yutaka Orihara at Tokyo University reported that the aromatic ring of salidroside and related compounds (acteosides) could be biosynthesized from Tyr through DA in olive leaf culture [72].

Thus, it is conceivable that reverse metabolism of salidroside could produce DA in mammals, making them feel very happy.

Icariin is a major flavonol in the herb epimedium that upregulates PTEN, the tumor-suppressing PIP3 phosphatase, eventually blocking PAK1 and extending significantly the lifespan of *C. elegans* [73,74]. Furthermore, Icariin has been known to display an antidepressant-like action on both behavioral despair and chronic mild stress models of mice. In 2011, a Chinese group at Fudan University in Shanghai, using a chronic social defeat protocol as a mouse model for depression, found that Icariin (25–50 mg/kg daily for a month) produces a remarkable increase in social interaction time by upregulating both affinity and expression of glucocorticoid receptor (GR), which is dramatically impaired in socially defeated mice [75]. Interestingly, GR is known to block the oncogenic kinase ERK, a downstream effector of PAK1 [76]. In other words, it is also conceivable that depressed animals with an impaired GR are predisposed to suffer from cancers or other PAK1/ERK-dependent diseases and die prematurely.

These findings together strongly indicate that PAK1 normally suppresses DA production or GR function, making people depressed, although the detailed molecular mechanism underlying these complex phenomena still remains to be clarified further in the future.

Since PAK blockers are able to boost DA in brains of mammals, as well as in insects such as bees, it would be reasonable to assume that bees feel very happy when they bump into anti-PAK compounds such as CAPE, apigenin, ARC, and triterpenes in the young buds of poplar trees, willow, camomile, and a few other trees or plants, and keep sucking these anti-PAK compounds to prepare propolis for the protection of their larvae.

Why does not propolis contain curcumin, berberine or salidroside? The answer is quite simple. Unlike CAPE and apigenin, these anti-PAK compounds are found mainly in roots hidden underground.

It should be pointed out that DA boosting is not the only way to suppress depression. CA (4 mg/kg) and its dimer rosmarinic acid (RA, 2 mg/kg) can suppress depression caused by FST in a mouse model, without boosting DA [77]. Like CAPE, both CA and RA block PAK1 but do not appear to boost DA.

References

[1] Allen J, Jaffer Z, Park SJ, Burgin S, et al. PAK1 regulates mast cell degranulation via effect on calcium mobilization and cytoskeletal dynamics. Blood 2009;113:2695–705.
[2] Lucanic M, Kiley M, Ashcroft N, L'etoile N, et al. The *C. elegans* PAKs are differentially required for UNC-6/netrin-mediated commissural motor axon guidance. Development 2006;133:4549–59.
[3] Maruta H. An innovated approach to in vivo screening for the major anti-cancer drugs Horizons in cancer research, 41: Nova Science Publishers; 2010. pp. 249–59.
[4] Maruta H, Burgess AW. Regulation of the Ras signalling network. Bioessays 1994;16:489–96.

[5] Hirokawa Y, Tikoo A, Huynh J, Utermark T, et al. A clue to the therapy of neurofibromatosis type 2: NF2/merlin is a PAK1 inhibitor. Cancer J 2004;10:20–6.

[6] Starinsky-Elbaz S, Faigenbloom L, Friedman E, Stein R, et al. The pre-GAP related domain of neurofibromin regulates cell migration through the LIM kinase/cofilin pathway. Mol Cell Neurosci 2009;42:278–87.

[7] Nur-E-Kmal MSA, Varga M, Maruta H. The GTPase-activating NF1 fragment of 91 amino acids reverses the v-Ha-RAS-induced malignant phenotype. J Biol Chem 1993;268:22331–22337.

[8] Tikoo A, Varga M, Ramesh V, Gusella J, et al. An anti-RAS function of neurofibromatosis type 2 gene product (NF2/Merlin). J Biol Chem 1994;269:23387–23390.

[9] Tang Y, Chen Z, Ambrose D, Liu J, et al. Kinase-deficient Pak1 mutants inhibit Ras transformation of Rat-1 fibroblasts. Mol Cell Biol 1997;17:4454–64.

[10] Tang Y, Marwaha S, Rutkowski J, Tennekoon G, et al. A role for Pak protein kinases in Schwann cell transformation. Proc Natl Acad Sci U S A 1998;95:5139–44.

[11] Hirokawa Y, Arnold M, Nakajima H, Zalcberg J, et al. Signal therapy of breast cancer xenograft in mice by the HDAC inhibitor FK228 that blocks the activation of PAK1 and abrogates the tamoxifen-resistance. Cancer Biol Ther 2005;4:956–60.

[12] Hirokawa Y, Nakajima H, Hanemann C, Kurtz A, et al. Signal therapy of NF1-deficient tumor xenograft in mice by the anti-PAK1 drug FK228. Cancer Biol Ther 2005;4:379–81.

[13] Grunberger D, Banerjee R, Eisinger K, Oltz E, et al. Preferential cytotoxicity on tumor cells by caffeic acid phenethyl ester isolated from propolis. Experientia 1988;44:230–2.

[14] Xu JW, Ikeda K, Kobayakawa A, Ikami T, et al. Down-regulation of Rac1 activation by caffeic acid in aortic smooth muscle cells. Life Sci 2005;76:2861–72.

[15] Matsuno T, Jung SK, Matsumoto Y, Saito M, et al. Preferential cytotoxicity to tumor cells of artepillin C isolated from propolis. Anticancer Res 1997;17:3565–8.

[16] Awale S, Li F, Onozuka H, Esumi H, et al. Constituents of Brazilian red propolis and their preferential cytotoxic activity against human pancreatic PANC-1 cancer cell line in nutrient-deprived condition. Bioorg Med Chem 2008;16:181–9.

[17] Demestre M, Messerli S, Celli N, Shahhossini M, et al. CAPE (caffeic acid phenethyl ester)-based propolis extract (Bio 30) suppresses the growth of human neurofibromatosis (NF) tumor xenografts in mice. Phytother Res 2009;23:226–30.

[18] Maruta H. Effective NF therapeutics blocking PAK1. Drug Disc Ther 2011;5:266–78.

[19] Messerli S, Ahn MR, Kunimasa K, Yanagihara M, et al. Artepillin C (ARC) in Brazilian green propolis selectively blocks the oncogenic PAK1 signaling and suppresses the growth of NF tumors in mice. Phytother Res 2009;23:423–7.

[20] Hodges A, Li S, Maynard J, et al. Pathological mutations in TSC1 and TSC2 disrupt the interaction between hamartin and tuberin. Hum Mol Genet 2001;10:2899–905.

[21] Manning B, Cantley L. Rheb fills a GAP between TSC and TOR. Trends Biochem Sci 2003;28:573–6.

[22] Saucedo L, Gao X, Chiarelli D, Li L, et al. Rheb promotes cell growth as a component of the insulin/TOR signalling network. Nat Cell Biol 2003;5:566–71.

[23] Kenerson H, Dundon T, Yeung RS. Effects of rapamycin in the Eker rat model of tuberous sclerosis complex. Pediatr Res 2005;57:67–75.

[24] Lee L, Sudentas P, Donohue B, Asrican K, et al. Efficacy of a rapamycin analog (CCI-779) and IFN-gamma in tuberous sclerosis mouse models. Genes Chromosomes Cancer 2005;42:213–27.

[25] Yalon M, et al. Regression of subependymal giant cell astrocytomas (SEGAs) with RAD001 (Everolimus) in TSC. Childs Nerv Syst 2011;27:179–81.

[26] Hara K, Maruki Y, Long X, Yoshino K, et al. Raptor, a binding partner of target of rapamycin (TOR), mediates TOR action. Cell 2002;110:177–89.

[27] Beevers C, Chen L, Liu L, Luo Y, et al. Curcumin disrupts the mammalian target of rapamycin-raptor complex. Cancer Res 2009;69:1000–8.

[28] Bharal N, Sahaya K, Jain S, Mediratta P, Sharma K. Curcumin has anti-convulsant activity on increasing current electroshock seizures in mice. Phytother Res 2008;22:1660–4.

[29] Blaheta R, Cinatl Jr J. Anti-tumor mechanisms of valproate: a novel role for an old drug. Med Res Rev 2002;22:492–511.

[30] Evason K, Collins J, Huang C, Hughes S, et al. Valproic acid extends C. elegans lifespan. Aging Cell 2008;7:305–17.

[31] Spannhoff A, Kim YK, Raynal N, Gharibyan V, et al. HDAC inhibitor activity in royal jelly might facilitate caste switching in bees. EMBO Rep 2011;12:238–43.

[32] Honda Y, Fujita Y, Maruyama H, Araki Y, et al. Lifespan-extending effects of royal jelly and its related substances on the nematode C. elegans. PLoS One 2011;6:e23527.

[33] Lee WH, et al. Human retinoblastoma susceptibility gene: cloning, identification and sequence. Science 1987;235:1394–9.

[34] Shan B, Durfee T, Lee WH. Disruption of RB/E2F-1 interaction by single point mutation in E2F-1 enhances S-phase entry and apoptosis. Proc Natl Acad Sci USA 1996;93:679–84.

[35] Leng X, Connell-Crowley L, Goodrich D, Harper J. S-phase entry upon ectopic expression of G1 cyclin-dependent kinases in the absence of retinoblastoma protein phosphorylation. Curr Biol 1997;7:709–12.

[36] Sherr C. Growth factor-regulated G1 cyclins. Stem Cells 1994;12(Suppl. 1):47–55.

[37] Nheu T, He H, Hirokawa Y, Walker F, et al. PAK is essential for RAS-induced up-regulation of cyclin D1 during the G1 to S transition. Cell Cycle 2004;3:71–4.

[38] Karasawa Y, Okisaka S. Inhibition of histone deacetylation by butyrate induces morphological changes in Y79 retinoblastoma cells. Jpn J Ophthalmol 2004;48:542–51.

[39] Bögler O, Huang H, Kleihues P, Cavenee W. The p53 gene and its role in human brain tumors. Glia 1995;15:308–27.

[40] Parada L, Land H, Weinberg R, Wolf D, et al. Cooperation between gene encoding p53 tumour antigen and ras in cellular transformation. Nature 1984;312:649–51.

[41] Li J, Yen C, Liaw D, Podsypanina K, et al. PTEN, a putative protein tyrosine phosphatase gene mutated in human brain, breast, and prostate cancer. Science 1997:1943–7.

[42] Wong AJ, Bigner S, Bigner D, Kinzler K, et al. Increased expression of the EGFR gene in malignant gliomas is invariably associated with gene amplification. Proc Natl Acad Sci U S A 1987;84:6899–903.

[43] Sawa H, Murakami H, Kumagai M, Nakasako M, et al. HDAC inhibitor, FK228, induces apoptosis and suppresses cell proliferation of human glioblastoma cells in vitro and in vivo. Acta Neuropathol 2004;107:523–31.

[44] Lee J, Lund-Smith C, Borboa A, Gonzalez A, Baird A, Eliceiri B. Glioma-induced remodeling of the neurovascular unit. Brain Res 2009;1288:125–34.

[45] Kuo HC, Kuo WH, Lee YJ, Lee WL, et al. Inhibitory effect of caffeic acid phenethyl ester on the growth of C6 glioma cells in vitro and in vivo. Cancer Lett 2006;234:199–208.

[46] Aoki H, Takada Y, Kondo S, Sawaya R, et al. Evidence that curcumin suppresses the growth of malignant gliomas in vitro and in vivo through induction of autophagy: role of AKT and ERK signaling pathways. Mol Pharmacol 2007;72:29–39.

[47] Wong M. Mammalian target of rapamycin (mTOR) inhibition as a potential antiepileptogenic therapy: from tuberous sclerosis to common acquired epilepsies. Epilepsia 2010;51:27–36.

[48] Okuma T, Kishimoto A. A history of investigation on the mood stabilizing effect of carbamazepine in Japan. *Psychiatry Clin Neurosci* 1998;52:3–12.

[49] Perucca E. Pharmacological and therapeutic properties of valproate: a summary after 35 years of clinical experience. CNS Drugs 2002;16:695–714.

[50] Beutler A, Li SD, Nicol R, Walsh M. Carbamazepine is an inhibitor of HDAC. Life Sci 2005;76:3107–15.

[51] Leung HT, Ring H. Epilepsy in four genetically determined syndromes of intellectual disability. J Intellect Disabil Res 2011 in press.

[52] Hayashi ML, Rao BS, Seo JS, Choi HS, et al. Inhibition of p21-activated kinase rescues symptoms of fragile X syndrome in mice. Proc Natl Acad Sci U S A 2007;104: 11489–11494.

[53] Eldridge. R, Denckla M, Bien E, Myers. S, et al. Neurofibromatosis type 1 (Recklinghausen's disease). Neurologic and cognitive assessment with sibling controls. Am J Dis Child 1989;143:833–7.

[54] Parker S, Mai C, Canfield M, Rickard R, et al. Updated national birth prevalence estimates for selected birth defects in the United States, 2004–2006. Birth Defects Res A Clin Mol Teratol 2010;88:1008–16.

[55]. Hattori M, Fujiyama T, Taylor H, Watanabe T, et al. The DNA sequence of human chromosome 21. Nature 2000;405:311–9.

[56] Delabar J, Aflalo-Rattenbac R, Créau N. Developmental defects in trisomy 21and mouse models. Sci World J 2006;6:1945–64.

[57] Troca-Marín J, Alves-Sampaio A, Montesinos M. An increase in basal BDNF provokes hyperactivation of the Akt-TOR pathway and deregulation of local dendritic translation in a mouse model of Down's syndrome. J Neurosci 2011;31:9445–55.

[58] Zhang H, Webb D, Asmussen H, Niu S, et al. A GIT1/PIX/Rac/PAK signaling module regulates spine morphogenesis and synapse formation through MIILC. J Neurosci 2005;25:3379–88.

[59] Castets M, Schaeffer C, Bechara E, Schenck A, Khandjian EW, Luche S, et al. FMRP interferes with the Rac1 pathway and controls actin cytoskeleton dynamics in marine fibroblasts. Hum Mol Genet 2005;14:835–44.

[60] Nasar S.. In: A beautiful mind. New York, NY: Simon and Schuster; 1998.

[61] Westfall R.. In: The life of Isaac Newton. Cambridge: Cambridge University Press; 1993.

[62] Bleuler M. Eugen Bleuler and schizophrenia. Brit J Psychiatry 1984;144:327–8.

[63] Chen SY, Huang PH, Cheng HJ. DISC1-mediated axon guidance involves TRIO–RAC–PAK small GTPase pathway signaling. Proc Natl Acad Sci U S A 2011;108:5861–6.

[64] Chubb J, Bradshaw N, Soares D, Porteous D, et al. DISC locus in psychiatric illness. Mol Psychiatry 2008;13:36–64.

[65] Kulkarni S, Dhir A. Berberine: a plant alkaloid with therapeutic potential for central nervous system disorders. Phytother Res 2010;24:317–24.

[66] Dickerson J, Sharp F. Atypical antipsychotics and a Src kinase inhibitor (PP1) prevent cortical injury produced by the psychomimetic, noncompetitive NMDA receptor antagonist MK-801. Neuropsychopharmacology 2006;31:1420–30.

[67] Boora K, Chiappone K, Dubovsky S, Xu J. Ziprasidone-induced spontaneous orgasm. J Psychopharmacol 2010;24:947–8.

[68]	Nakazawa T, et al. Anti-depressant effects of apigenin (and TMCA) in the forced swimming test. Biol Pharm Bull 2003;26:474–80.

[69]	Xu Y, Ku BS, Yao HY, Lin YH, et al. The effects of curcumin on depressive-like behaviors in mice. Eur J Pharmacol 2005;518:40–6.

[70]	Penga WH, Lo KL, Lee YH, Hung TH, et al. Berberine produces antidepressant-like effects in the forced swim test and in the tail suspension test in mice. Life Sci 2007;81:933–8.

[71]	Panossian A, et al. Comparative studies of Rhodiola rosea preparations on behavioral despair of rats. Phytomedicine 2008;15:84–91.

[72]	Saimaru H, Orihara Y. Biosynthesis of acteoside in cultured cells of *Olea europaea*. J Nat Med 2010;64:139–45.

[73]	Kim SH, Ahn KS, Jeong SJ, Kwon TR, et al. Janus activated kinase 2/signal transducer and activator of transcription 3 pathway mediates icariside II-induced apoptosis in U266 multiple myeloma cells. Eur J Pharmacol 2011;654:10–16.

[74]	Cai WJ, Huang JH, Zhang SQ, Wu B, et al. Icariin and its derivative icariside II extend healthspan via insulin/IGF-1 pathway in *C. elegans*. PLoS One 2011;6:e28835.

[75]	Wu JF, Du J, Xu CQ, Le JJ. Icariin attenuates social defeat-induced down-regulation of glucocorticoid receptor in mice. Pharmacol Biochem Behav 2011;98:273–8.

[76]	Kharwanlang B, Sharma R. Molecular interaction between the glucocorticoid receptor and MAPK signaling pathway: a novel link in modulating the anti-inflammatory role of glucocorticoids. Indian J Biochem Biophys 2011;48:236–42.

[77]	Takeda H, Tsuji M, Inazu M, Egashira T, et al. Rosmarinic acid and caffeic acid produce antidepressive-like effect in the forced swimming test in mice. Eur J Pharmacol 2002;449:261–7.

6 PAK1 in Alzheimer's and Huntington's Diseases

Qiu-Lan Ma[1,2], Fusheng Yang[1,2], Sally A. Frautschy[1,2], and Greg M. Cole[1,2]

[1]Department of Neurology, University of California, Los Angeles, [2]Geriatric Research and Clinical Center, Greater Los Angeles Veterans Affairs Healthcare System, West Los Angeles Medical Center, Los Angeles, CA, USA

Abbreviations

Aβ	amyloid-β
AD	Alzheimer's disease
ADDLs	β-amyloid-derived oligomers
APP	amyloid precursor protein
BBB	blood–brain barrier
CA1	Corpus Ammon 1
CRMP-2	collapsin response mediator protein 2
CTCL	cutaneous T-cell lymphoma
DHA	docosahexaenoic acid
HD	Huntington's disease
HDAC	histone deacetylase
LIMK	LIM kinase
LTP	long-term potentiation
MR	mental retardation
NFT	neurofibrillary tangles
NMDARs	N-methyl-D-aspartate receptors
PAK	p21-activated kinase
PSD	postsynaptic density

6.1 Alzheimer's Disease

6.1.1 Introduction

Alzheimer's disease (AD) is the most common form of dementia. So far there is no cure for this disease, which eventually leads to death. It was first described in

PAKs, RAC/CDC42 (p21)-activated Kinases. DOI: http://dx.doi.org/10.1016/B978-0-12-407198-8.00006-0

Figure 6.1 Alois Alzheimer (1864–1915).

1906 by German psychiatrist and neuropathologist Alois Alzheimer (1864–1915, see Figure 6.1), and thus named after him. Most often, AD is diagnosed in people over 65 years of age although AD can occur much earlier (this is less prevalent). In 1989, Evans and collaborators [1] showed that of those over 65, more than 10% had AD. This prevalence rate was strongly associated with age: of those 65–74 years old, only 3% had AD, compared with nearly 20% of those 75–84 years old and nearly 50% of those over 85. In 2006, there were around 30 million AD sufferers worldwide. AD is now predicted to affect more than 1% of the world population by 2050.

AD is a prevalent neurodegenerative disease characterized clinically by progressive cognitive decline with aging, and pathologically by decades-long prodromal accumulation of neuritic plaques containing amyloid-β (Aβ) protein and neurofibrillary tangles (NFT) containing tau protein aggregates of paired helical filaments. Proximal to detectable cognitive decline, there is a region-dependent selective loss of synapses, especially excitatory synapses and vulnerable neurons. Selective region-dependent synaptic loss is generally observed at early stages of AD clinical symptoms and is more closely related to cognitive deficits than neuronal loss or amyloid buildup [2,3]. Recently, soluble aggregated Aβ forms called β-amyloid-derived diffusible ligands (ADDLs) or Aβ oligomers, including dimers, trimers and dodecamers (12-mer or Aβ *56) have been proposed as important neurotoxic forms of Aβ responsible for synaptic dysfunction and loss in AD patients and animal models [4–7]. Anti-Aβ antibody-mediated reduction of Aβ oligomers can rescue synaptic deficits in AD animal models [8,9]. However, these monoclonal antibodies cross the blood–brain barrier (BBB) poorly, and typically their oligomer binding competes with larger pools of plaque and vascular amyloid deposits, and therefore, their therapeutic effect may be rather limited clinically. Thus, in an attempt to develop far more effective (BBB-permeable) anti-AD drugs, identification of the major signal transduction pathways responsible for Aβ- and tau-dependent synaptic deficits are being aggressively pursued.

Synaptic plasticity in general is dependent on the regulation of the actin-based cytoskeleton in dendritic spines [10–12]. The regulation of actin filament (F-actin) dynamics involves various actin-binding proteins and their binding partners, including membrane receptors and their downstream signaling cascades. In particular,

Rho family GTPases (RHO, RAC, and CDC42) play a central role in regulating actin reorganization through a variety of their downstream effectors, such as PAKs, ROCK, and WASPs (Wiskott–Aldrich syndrome proteins, also known as CDC42-dependent actin-severing proteins) [13–15]. RHOA and RAC/CDC42 are linked to distinct upstream and downstream signaling pathways to promote morphogenesis of dendritic spines via their antagonistic effects on synaptic plasticity. RHOA activates the kinase ROCK which promotes amyloid precursor protein (APP) processing to its derivative putative toxic species, Aβ42 [16]. The RAC/CDC42-activated kinase PAK1 is a key regulator for actin cytoskeleton and dendritic spine morphogenesis. Several years ago we found a loss of PAK1 in the cytoplasm of AD brain specimens, as well as AD animal and cellular models, suggesting PAKs might play crucial roles in dendritic spine/synapses loss and cognitive defects in AD [17]. Here we shall focus our discussion on the pathologic alterations of PAK1 and its downstream LIM kinase (LIMK) signaling pathways, with an emphasis on the potential therapeutic value of PAK1 blockers for AD.

6.1.2   Alteration of PAKs in AD

Cognitive decline has been directly linked to synaptic dysfunction, especially loss of synapses, postsynaptic receptor complex components and dendritic spines in AD and mental retardation (MR) syndromes [2,18]. A primary role for dendritic spine defects in MR has come from identification of mutations in X-linked MR genes [18]. These MR genes code for a set of proteins implicated in the postsynaptic pathways regulating the dynamics of dendritic spine actin assembly and disassembly and spine morphogenesis as well as synapse formation and stability. PAK3 is one of these MR genes, and missense mutation in PAK3 causes severe X-linked nonspecific MR [19,20]. Animal models of MR syndromes using the autoinhibitory domain of PAK1, which blocks PAK1–3 [21] or knockout of its downstream LIMK, show defects in dendritic spines and cognitive function [22]. These suggest an essential role for the PAK1–3/LIMK pathway in regulating synaptic plasticity.

Similar to their occurrence in MR, postsynaptic and dendritic spine defects occur early in AD, where they are spatiotemporally situated to play a critical role in cognitive deficits. For example, although the overall estimate of neuronal loss in AD hippocampus is around 5–40% depending on stage, overall loss of postsynaptic proteins from whole hippocampus, such as actin-regulating developmentally regulated brain protein (drebrin), are reported to reach 70–95% [23–25]. Drebrin regulates actin dynamics and is involved in postsynaptic function in excitatory synapses. For example, drebrin-dependent actin filaments in dendritic filopodia govern synaptic targeting of PSD-95 and dendritic spine morphogenesis [26].

Higher-than-normal drebrin levels also have a significant impact. Andrew Matus's group at the Friedrich Miescher Institute (Basel, Switzerland) found that upregulation of drebrin in rat hippocampal neurons destabilized mature dendritic spines so that they lost synaptic contacts and came to resemble immature dendritic filopodia [27]. Furthermore, this drebrin-induced spine destabilization was dependent on RAS activation: oncogenic RAS mutant, which activates the PAK1-LIMK pathway, destabilized spine morphology, whereas drebrin-induced spine destabilization was rescued by a dominant-negative RAS mutant. Conversely, RNAi-mediated drebrin knockout prevented RAS-induced destabilization and promoted spine maturation in developing neurons. Thus, it still remains unclear if the observed downregulation (or loss) of drebrin in AD is just a consequence of the AD pathogenesis or is among the key causes of AD. Furthermore, it still remains to be clarified how debrin is selectively lost in AD. Since drebrin is enriched in calpain-sensitive PEST (Pro, Glu, Ser, Thr-rich) sequences, its loss may reflect Aβ oligomer-induced excitotoxicity.

Two major neuronal isoforms of PAKs exist in the brain, PAK1 and PAK3. Normally, both forms have a diffuse distribution in cell bodies and dendrites. In contrast, in AD brain, significant losses of total PAK1 (around 35%) and PAK3 (55–69%) were observed in hippocampus, while total PAK3 was also significantly decreased in AD temporal cortex (63–77%). However, the loss of PAK1's kinase activity clearly exceeds that of its protein level. The active autophosphorylated PAK1 at Ser 141 had a more severe 73% decrease in AD temporal cortex [17]. In addition, PAK1 also showed aberrant activation and the activated phospho-PAK1 (pPAK) was translocated from cytoplasmic to membrane fractions, most obviously to granular structures in a complex with its activators, GTPases RAC/CDC42 [9]. Alterations in PAK are stage dependent with an increase in the total protein level of PAK1/3 at early stages of AD but a reduction of both total and cytoplasmic pPAK in late-stage severe AD reported by Nguyen et al. [28]. Because a similar reduction and subcellular translocation of PAK1 was observed in a transgenic Aβ aggregate-accumulating mouse model with limited neurodegeneration [9], these changes may be downstream from Aβ aggregates, and may not simply reflect a loss of neuronal compartments. Similarly, cytoplasmic pPAK was significantly reduced in a triple transgenic mouse model of AD that develops both plaque and tangle pathology, and both this loss of pPAK and cognitive deficits were ameliorated by a diet enriched in docosahexaenoic acid (DHA) [29]. Since PAK1–3 defects are sufficient to cause cognitive deficits, these data suggest that deregulation (hyperactivation) of PAK1 may also play an important role in dendritic spine/synapse and memory defects in AD.

To characterize potential mechanisms contributing to cytoplasmic PAK deficits in AD, we cultured hippocampal neurons with β-amyloid oligomers and observed an initial rapid abnormal PAK activation and translocation followed by a delayed loss of cytoplasmic pPAK. This aberrant activation was accompanied by a rapid loss of F-actin and dendritic spines, unlike the response to normal activation of PAK1 by RAC/CDC. The wild-type (WT) PAK1, but not its kinase-dead mutant (K299A), prevented pathological changes in spines, providing evidence that useful PAK1 recruitment and signaling is blocked by β-amyloid exposure [9]. One possibility is that Aβ activation of extrasynaptic NMDA receptors differs radically from synaptic NMDA receptor activation of RAC/PAK, which leads to dendritic arbor growth.

6.1.3   The RAC/PAK1/LIMK/Cofilin Signaling Pathway in AD

In 2007, Ricardo Maccioni's group [30] at the University of Chile (Santiago, Chile) found that the enhancement in actin polymerization induced by fibrillar amyloid-β

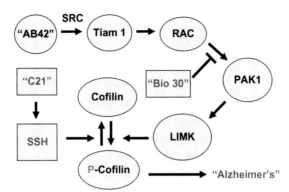

Figure 6.2 *A PAK-related pathway leading to AD and its intervention.* Aβ42-induced activation of LIMK, essential for the AD, could be blocked by either compound 21 (C21), which eventually activates slingshot (SSH), a cofilin phosphatase, or Bio 30, which inactivates PAK1, which is essential for LIMK activation.

peptide (Aβ) is associated with an increased activity of GTPases, RAC, and CDC42. RAC upregulation involves the participation of Tiam1, an RAC guanine-nucleotide exchange factor; Aβ exposure leads to Tiam1 activation by a Ca^{2+}-dependent mechanism (Figure 6.2). Furthermore, they reported that fibrillar Aβ42 (fAβ) treatment of hippocampal neurons can also activate PAK1 [31]. These results suggest that β-amyloid species could be the primary elements responsible for PAK dysfunction in AD. PAK1 regulates actin cytoskeletal dynamics through LIMK, which phosphorylates cofilin at Ser 3 and inactivates its F-actin severing activity [32].

Is this PAK1-LIMK-cofilin signaling pathway involved in AD-associated pathogenesis? Pathologic intracellular inclusion bodies (Hirano bodies) containing cofilin-decorated actin rods and other actin-binding proteins are prominent features in the hippocampus and cortex of AD brains [33,34]. We observed that confocal co-labeling of pPAK and cofilin in AD hippocampus reveals cells with different stages of pPAK and cofilin pathologies; for instance, some cells exhibit increasingly intense cofilin labeling associated with progressively decreased diffuse pPAK but with granular structure staining (Figure 6.3). The severe pPAK and cofilin pathologies in AD are associated with a reduction in the dendritic spine drebrin [24–26]. This is consistent with the hypothesis that loss of the cytosolic pPAK can lead to local pathology related to cofilin aggregation, drebrin loss, and synaptic defects observed in AD brain [17]. However, it still remains to be clarified how severe debrin loss affects the actin cytoskeleton in AD brain.

Both β-amyloid oligomers and fibrillar Aβ42 (fAβ) treatment of hippocampal neurons can activate PAK1, which in turn activates LIMK *in vitro* [17,31].

Alfredo Lorenzo's group in Argentina found that fAβ activates LIMK and induces cofilin phosphorylation in cultured neurons. Furthermore, in AD brain, the number of pLIMK-positive neurons was significantly increased in those regions affected with AD pathology. Further, S3, a synthetic peptide, which blocks the cofilin phosphorylation by LIMK, inhibited fAβ-induced neuronal degeneration,

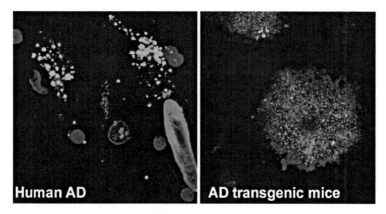

Figure 6.3 *Confocal co-labeling of pPAK and cofilin in AD hippocampus.* Some neurons exhibit intense cofilin labeling (red) and granular pPAK staining. Blue color represents DAPI. (For interpretation of the references to color in this figure legend, the reader is referred to the web version of this book.)

clearly indicating the involvement of LIMK in Aβ-induced neuronal degeneration *in vitro* [35].

One important question remaining is whether the kinase ROCK, which also inactivates cofilin through LIMK, is involved in the fAβ-induced neurotoxicity. Although direct ROCK inhibitors such as Y-27632 and H-1152 induce a rapid cofilin dephosphorylation (activation) and robust neurite outgrowth of PC12 cells [36], so far there is no solid evidence proving that such a ROCK-specific inhibition alone affects fAβ-induced neurotoxicity.

6.1.4   Connections of NMDA Receptors, FYN, and PAK Signaling in AD

N-methyl-D-aspartate receptors (NMDARs) are a subtype of ionotropic glutamate receptor that play a central role in synaptic mechanisms of learning and memory [37]. NMDARs are directly anchored to postsynaptic density (PSD) and thus to the actin cytoskeleton. They flux Ca^{2+} and are critical for dendrite spine formation induced by neuronal activity. Recent studies have shown that Ca^{2+} influx can stimulate the CaMKK/CaMKI cascade, which activates GIT1, an RAC GEF that leads to downstream PAK activation involved in synaptogenesis [38]. Thus, crosstalk between NMDARs and Rho family GTPases via calcium signaling to activate RAC/PAK plays an important role in the synaptic plasticity underlying new synapse formation.

Interestingly, however, messenger RNA levels for NMDAR subunits NR2A and NR2B are decreased in AD hippocampus and entorhinal cortex [39]. Theoretically, extensive neuron loss could contribute to this but it has also been reported that decreased protein subunits of NMDAR-associated proteins (e.g., NR2B as well as the scaffold PSD-95 and activated alpha-CaMKII) occur in PSD preparations from APP[V717I] AD transgenic mice that lack extensive neuron loss. This NMDAR

component loss is associated with impaired NMDA-dependent long-term potentiation (LTP), a major cellular mechanism implicated in learning and memory and
decreased NMDA- and AMPA-receptor currents in the hippocampal CA1 (comu
ammonis 1) region [40]. NMDA receptors also mediate Aβ oligomer-induced effects
on dendritic spine and synaptic marker loss in cultured neurons [41,42].

The SRC family Tyr kinase FYN associates with the NMDAR complex and phosphorylates NMDA receptor subunits NR2A and NR2B. Robust phosphorylation of
NR2B at Tyr1472 by FYN has been implicated in LTP [43]. We found that Aβ oligomers decreased pNR2B Tyr1472 levels in both membrane and cytosol fractions
without altering the membrane/cytosol ratio [9]. The SRC family Tyr kinase inhibitor PP2 does not block these oligomer effects, but significantly blocks both RAC and
pPAK translocation in Aβ oligomer-treated primary neurons, indicating FYN activation is independent of the Aβ-induced pNR2B loss/dephosphorylation. Thus, Aβ oligomer-induced RAC/PAK changes appear downstream from SRC/FYN [9]. In fact,
PP1, a PP2 derivative, blocks both RAS-induced PAK1 activation and growth of RAS-
induced transformants (cancers) *in vivo* without any adverse effect [44]. Although
the responsible direct target of PP1 among SRC family kinases remains to be clarified, FYN appears to be among its likely targets, because the IC_{50} of PP1 for blocking
PAK1 is around 10 nM, which is very close to PP1's IC_{50} for FYN [45]. Intriguingly,
oncogenic RAS (upstream of PAK1) has been found to upregulate FYN mRNA dramatically (>100-fold) while both siRNA for FYN and the inhibitor PP2 strongly
inhibit the PAK1-dependent metastasis and invasion of human breast cancers [46].

FYN has been implicated in soluble oligomer (ADDLs) induced LTP defects *in
vitro*, and in synaptotoxicity and cognitive deficits in APP transgenic mice. Collectively,
these data suggest that rapid abnormal FYN-dependent activation and translocation of
RAC/PAK1 is likely to contribute to synaptic dysfunction and excitatory synaptic deficits involved in a pathway involving Aβ oligomers, NMDARs, and FYN in AD.

6.1.5 *Amyloidβ-Induced Microglial Phagocytosis of Neurons in AD*

Amyloid can also activate monocytic lineage cells and induce phagocytosis which is
regulated by many of the same players, including. src family kinases, CDC42/ Rac/
PAK/LIMK/cofilin [47]. It is known that high concentrations of Aβ (µM) can induce
direct toxicity in neurons. However, Guy Brown's group [48] at the University of
Cambridge in the United Kingdom recently found that low concentrations of Aβ
(nM) can also induce neuronal loss through a microglia-mediated mechanism. In
mixed neuronal-glial cultures from rat cerebellum, 250 nM Aβ1–42 (added as monomers, oligomers, or fibers) induced about 30% loss of neurons between 2 and 3 days.
This neuronal loss occurred without any increase in neuronal apoptosis or necro-
sis, and no neuronal loss occurred with Aβ42–1. In this model, Aβ greatly increased
the phagocytic activity of microglia and induced phosphatidylserine (PS) exposure
on neuronal processes, a signal used to promote phagocytosis of apoptotic cells or
cell fragments. Blocking exposed PS by adding annexin V or an antibody to PS, or
inhibiting microglial phagocytosis by adding either cytochalasin D (1 µM to block
actin polymerization) or cyclo(RGDfV) to block vitronectin receptors, significantly

prevented neuronal loss. Loss of neuronal synapses occurred in parallel with loss of cell bodies and was also prevented by blocking phagocytosis. Inhibition of phagocytosis prevented neuronal loss with no increase in neuronal death, even after 7 days, suggesting that microglial phagocytosis was the primary cause of neuronal death induced by nanomolar Aβ. Of course, cytochalasin D would not be a selective AD therapeutic.

6.1.6   PAK1 Inhibitors or Blockers for AD Therapy

As discussed in detail in Chapter 3, full activation of PAK1 requires not only RAC/CDC42 and a few SH3 adaptor proteins such as PAK-interacting exchange factor (PIX) and NCK, but also at least two distinct Tyr kinases. One is FYN, the target of PP1/PP2 as we discussed above. The other is ETK, which directly phosphorylates PAK1. The ETK-PAK1 interaction is selectively blocked by AG 879 or its water-soluble derivative GL-2002/GL-2003, which has an IC_{50} of 5–10 nM [49]. The combination of PP1 and AG 879 or GL-2003 (20 mg/kg, i.p., twice a week) completely blocks RAS-induced PAK1 activation and PAK1-dependent growth of RAS cancers, such as human pancreatic and colon cancers *in vivo* (xenografts in mice), without any detectable adverse effects [50]. Both PP1 and AG 879 derivatives pass the blood–brain barrier (BBB). Thus, it is conceivable that this combination of FYN and ETK inhibitors is a useful candidate for the treatment of AD.

As discussed in Chapter 3, several PAK inhibitors such as TAT-PAK18, CEP-1347, UnPAK309, and FK228 have been developed for the therapy of mainly non-brain solid tumors. These inhibitors block the growth of PAK1-dependent solid tumor cells selectively without affecting the growth of normal cells [51–54]. However, the most potent inhibitor, FK228 (IC_{50} around 1 nM), fails to pass the BBB.

A decade ago, a group at Cephalon found that CEP-1347, which somehow blocks the activation of the kinase JNK, suppresses Aβ42-induced neuronal cell death [55]. Later CEP-1347 was identified as a direct inhibitor of RAC/CDC42-dependent kinases such as PAK1–3 and MLKs, which activate JNK [53], suggesting that either PAKs or MLKs are essential for AD pathogenesis. Then, around 2006, an Argentine group led by Alfredo Lorenzo [35] found that the LIMK, just downstream of PAK1, is required for the neurotoxicity of amyloid-beta in cell culture. Phosphorylated LIMK is increased in the brains of AD patients compared with those of unaffected (healthy) individuals. LIMK is phosphorylated at Thr 508 by PAK1 (but not by MLK) for activation. These observations strongly suggested that CEP-1347 suppresses Aβ42-induced neuronal cell death, mainly by blocking the PAK1-LIMK signaling pathway. Nevertheless, since the IC_{50} of both CEP-1347 and TAT-PAK18 are high (around or above 1 μM) their therapeutic potential may be limited. In fact, clinical trials of CEP-1347 for Parkinson's disease (PD) were terminated during phase 2 around 2004.

More recently, potent PAK1-specific inhibitors with IC_{50} around 10 nM were developed by Afraxis in San Diego for therapy of PAK1-dependent CNS diseases, including AD and Huntington's disease (HD). These PAK1-inhibitors do not inhibit PAK4–6, and pass the BBB. Thus, these newly developed potent PAK1-specific inhibitors could potentially serve as possible AD therapeutics.

While aberrant hyperactivation of PAK is driven by Aβ, PAK signaling plays a critical role in synaptic plasticity, learning, and memory, suggesting a limited therapeutic window for inhibiting PAK1 as a direct drug target for AD or other brain diseases. Despite this caveat, based on the dynamic alteration of PAK1 in different AD stages, PAK1 inhibitors might be still useful for AD intervention to block abnormal PAK1 activation and translocation. One can also target the upstream Aβ oligomers. As discussed in Chapter 3, curcumin is a natural direct PAK1 inhibitor that can pass the BBB. We found that curcumin significantly suppressed persistent phospho-PAK translocation to granules in CA1 neurons evaluated in aged APPswe Tg2576 mice. Curcumin also suppressed punctuated anti-Aβ staining and pPAK translocation induced by Aβ42 oligomers in cultured hippocampal neurons [9]. Since curcumin has been reported as an effective anti-amyloid and anti-Aβ oligomer agent *in vivo* and *in vitro* [56,57], curcumin's activity is likely through the reduction of upstream amyloid, notably oligomeric aggregates. However, curcumin alone has a very poor bioavailability, which has contributed to failed clinical trials; it clearly has to be encapsulated with liposomes or otherwise formulated for good oral absorption in the clinic [58].

In 2011, Maccione's group found that AB42 oligomer activates LIMK through RAC/CDC42 and PAK1, leading to the inactivation of cofilin [31]. Furthermore, a cofilin phosphatase called SSH, which antagonizes LIMK, blocked the neurocytotoxicity of AB42 oligomer [31]. In short, AB42 oligomer blocks cofilin's F-actin severing activity through the SRC-Tiam1-RAC/CDC42-PAK1-LIMK signaling pathway (Figure 6.2), and SSH could reverse this neurodegenerative pathway. Thus, in principle, SSH activator(s), in addition to a water-soluble (aminohexyl) derivative of SRC family kinase inhibitors such as PP1 and PP2, and PAK1/LIMK blockers, provide potential leads for oral therapy, particularly for early stages of AD.

In this context, it should be worth noting that stimulation of the angiotensin II type 2 (AT_2) receptor by a recently synthesized potent AT_2 receptor agonist called compound 21 (C21) enhanced cognitive functions (in particular spatial learning) that were impaired in an AD mouse model in which AB42 had been injected intracerebroventricularly [59]. This nonpeptide compound (Figure 6.4; the Km for AT_2 receptor is just below 1 nM) was developed in 2004 by a Swedish group led by Anders Hallberg and Mathias Alterman at Uppsala University [60,61]. How does this orally active compound work? Figure 6.2 shows that the C21 activates the SSH (just as angiotensin II does) via the AT_2 receptor by stimulating the formation of endogenous ROS via oxidation of 14-3-3zeta, increasing membrane ruffling and cell motility [62]. More interestingly, however, the C21 compound (1 mg/kg daily) appears to suppress LPS-induced inflammation by inhibiting NF-kappaB in hypertensive rats, but without affecting blood pressure [61]. C21, which may also inhibit PAK1, which is essential for inflammation, is currently in clinical trials (phase 1 testing) by a Swedish biotech company called Vicore Pharma.

Although neither the potent PAK1 blocker FK228 nor the PAK1 inhibitor CEP-1347 is available on the market for neurodegenerative diseases as yet, several natural PAK1 blockers are available on the market inexpensively. One of them is berberine, which has been used in traditional medicinal extracts and has pleiotropic activities. It

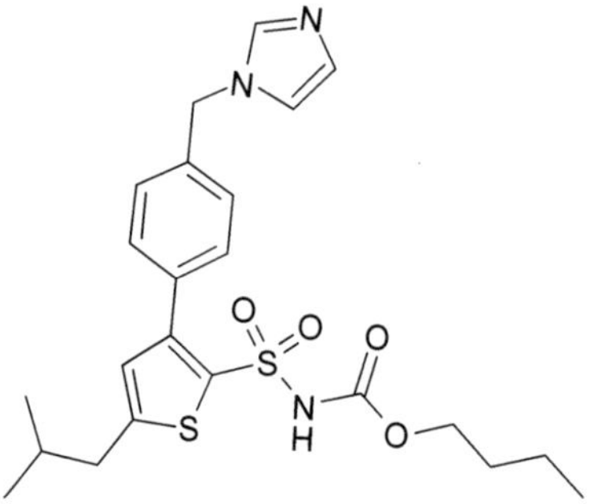

Figure 6.4 *Compound 21 (C21).*
This image is derived from a figure in
Ref. [57], kindly provided by
Dr. Mathias Alterman.

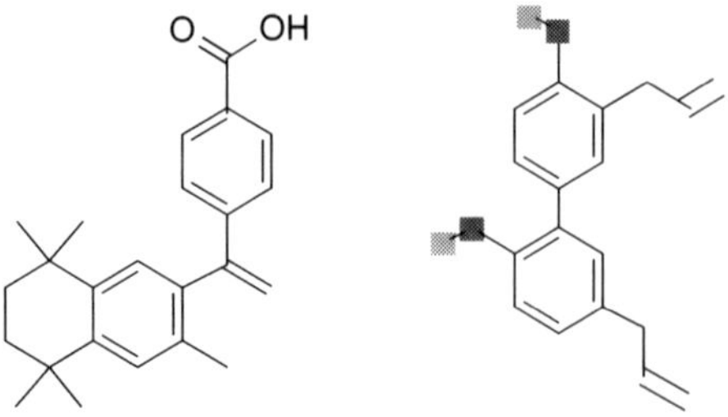

Figure 6.5 Bexarotene and honokiol (RXR agonists).

has been shown to ameliorate spatial memory impairment in a rat Aβ infusion model of AD [63]. Furthermore, a far more potent analog (diethyl substitute) was recently developed by Doug Kinghorn's group at Ohio State University [64]. Alternatively, as discussed in Chapter 5, propolis extracts such as Bio 30 are potentially useful for therapy of PAK1-dependent neurodegenerative diseases such as AD and HD.

Recently, Gary Landreth's team at Case Western Reserve University (Cleveland, OH) revealed a surprise discovery: a synthetic retinoid X receptor (RXR) agonist called bexarotene (Figure 6.5) rapidly clears Aβ42 aggregates and reverses the learning deficit in AD mice [65]. Like histone deacetylase (HDAC) inhibitors such as SAHA and FK228, this drug is currently available on the market for the therapy of a rare cancer, cutaneous T-cell lymphoma (CTCL), but the AD mouse work showing upregulation of ApoE needs to be repeated with longer term therapy and extended to include ApoE4 before moving to the clinic. Interestingly, a natural RXR agonist from magnolia bark extract called honokiol (Figure 6.5) blocks PAK1 by downregulating its direct activator RAC [66], an effector of Aβ oligomer toxicity. Furthermore, these two distinct RXR agonists work synergistically to activate the PPARγ/RXR heterodimers [67]. Thus, this combination or even honokiol alone might serve as an inexpensive AD/HD therapeutic in the future. The honokiol-based magnolia bark extract (300 mg/per capsule) is inexpensive (US ~$0.030), compared to bexarotene (75 mg/per capsule) (US ~$20.00). This magnolia bark extract is a Chinese traditional medicine that has been used for more than 2000 years to treat depression, epilepsy, inflammation, type 2 diabetes, and solid tumors, all of which are PAK1-dependent diseases, as discussed in Chapters 3 and 5.

6.2 Huntington's Disease

HD is an autosomal-dominant progressive neurodegenerative disorder that prominently affects the basal ganglia leading to clinically significant motor function, cognitive, and behavioral deficits. HD is caused by an expanded CAG repeat encoding a polyglutamine (polyQ) tract in exon 1 of the HD gene. Normal HD alleles have 37 or fewer glutamines in this polymorphic tract; having more than 37 of these residues causes HD [68]. A polyglutamine (polyQ) repeat expansion of more than 37 units, as observed in HD, results in a very large and aggregation-prone protein (>348 kDa) of 3145 or more amino acids called huntingtin (Htt) that accumulates as aggregates in HD. The length of the CAG tract is directly correlated with aggregation rate and disease onset, with longer expansions leading to earlier onset of HD. The onset age in HD patients with CAG repeats below 60 units varies considerably and may depend on variable endogenous defenses, including clearance mechanisms.

Although the hallmarks of HD are motor disability and chorea, while in contrast the main symptom of AD is dementia, HD and AD patients share many of the same clinical manifestations. These include behavioral and psychiatric disturbances such as depression and apathy in the initial stages of disease, and eventual cognitive defects that result in forgetfulness, impaired judgment, disorientation, and confusion. However, cognitive deficits in patients with HD are usually less severe than in AD.

The mechanisms for mutant Htt-induced cellular toxicity remain incompletely understood. Several studies report that WT Htt can limit the cellular toxicity of mutant huntingtin *in vitro* and *in vivo*; WT Htt protects neurons through inhibition of procaspase-9 processing or caspase 3-mediated pathways [69–72]. For example, WT Htt prevents the cleavage of PAK2 by caspase-3 and caspase-8, a modification that activates PAK2 by releasing a constitutively active C-terminal kinase domain that mediates cell death. Thus, overexpression of normal WT huntingtin significantly inhibits caspase-3-mediated and caspase-8-mediated cleavage of PAK2 in cells [73]. Therefore, it has been proposed that loss of function of mutant Htt might contribute to neuronal toxicity resulting from the polyQ expansion [74]. In contrast, genetic and transgenic data argue that the major toxic pathways activated by the mutation of the HD gene are via a gain of function caused by intracellular aggregates of mutant Htt protein. However, it remains unclear whether any of the toxic actions of mutant Htt are primarily due to loss of function of soluble monomers, or whether most toxic actions are gains of negative function with soluble oligomers or more insoluble species [75], In this respect, the same questions arise with AD and tau aggregates. And as with AD, PAK1 appears to modulate toxic pathways in HD.

6.2.1 PAK1 in HD Pathogenesis

Recently, PAK 1 was identified as an Htt interactor that modifies mutant Htt (muHtt) toxicity [76]. PAK1 promoted soluble mutant huntingtin self-interaction, which enhanced toxicity in HD cellular models [76], suggesting that PAK1 may play an important role in HD pathogenesis. PAK1 colocalized with muHtt aggregates in cell models and in human HD brains. PAK1 overexpression not only enhanced the aggregation of muHtt, but also promoted soluble WT Htt (WTHtt)–WTHtt, WTHtt–muHtt, and muHtt–muHtt interactions. Furthermore, PAK1 overexpression enhanced Htt toxicity in cell models and neurons in parallel with its ability to promote aggregation, while PAK1 knockout suppressed both aggregation and toxicity. Interestingly, overexpression of either kinase-dead or WT PAK enhanced both aggregation and toxicity of muHtt protein. The domains of PAK1 that bind Htt also facilitate oligomerization/ aggregation, and no enhanced toxicity was observed with PAK1 domains that do not bind Htt. More importantly, PAK1 also enhances dimerization of WTHtt but does not form any large or toxic aggregates. This suggests that PAK1 plays a key role in enhancing Htt–Htt interactions in a way that synergizes with the effects of the 'sticky' expanded polyQ tract to enhance aggregated muHtt toxicity [76].

Supporting an important role for a PAK pathway in muHtt toxicity, a PAK1–3 activator called PIX has also been identified as a novel huntingtin-interacting protein [77]. Similar to PAK1, PIX binds both the N-terminal region of WTHtt and muHtt, and colocalizes with muHtt in cells, where it accumulates in muHtt aggregates. Deletion analysis argues that the dbl homology and pleckstrin homology domains of PIX are required for PIX interaction with Htt. Increasing PIX expression enhanced muHtt aggregation by inducing SDS-soluble mutHtt–mutHtt interactions. Conversely, knocking out PIX attenuated muHtt aggregation [77]. These findings indicate that PIX plays an important role in muHtt aggregation, and inhibition of the PAK-PIX interaction could be a useful strategy for HD therapy.

6.2.2 PAK1 Inhibitors or Blockers for HD Therapy

In principle, candidate drugs or natural products that block or delay toxic polyQ aggregation or aggregate effectors could serve as useful HD therapeutics, even though they cannot cure HD. In 2004, a Japanese group at Riken found that a natural disaccharide called trehalose reduces polyQ aggregate accumulation, and significantly delays the onset of HD in a mouse model [78]. Trehalose promotes autophagy and removal of pathogenic aggregates in models of HD and AD [79,80].

Although the detailed molecular mechanisms underlying the anti-HD action of trehalose still remain to be more precisely studied, there is indirect evidence suggesting that trehalose serves as a chaperone, just like heat-shock proteins, which block the aggregation of misfolded proteins such as heat-denatured proteins [81]. C. albicans yeast cells growing exponentially on glucose are extremely sensitive to severe heat shock (52.5°C for 5 min). When these cultures were subjected to a moderately elevated temperature preincubation (42°C), they became thermotolerant and displayed higher resistance to further heat stress [81]. The intracellular content of trehalose was very low in exponential cells but underwent a marked increase upon this nonlethal conditioning heat exposure, whereas the external trehalose remained practically unmodified. The accumulation of trehalose is likely due to heat-induced activation of the trehalose-6-phosphate synthase (TPS) complex. This speculation is based on the recent observation by a Japanese group led by Yoko Honda that trehalose extends the lifespan and increases the heat resistance of C. elegans by activating the stress-resistant genes TPS-1 and TPS-2, which encode TPSs [82].

It is possible that facilitating a heat-shock response in HD could be useful for HD therapy. At least with AD in a nematode model, heat-shock treatment and induction of heat-shock proteins such as HSP16 in AD worms expressing beta-amyloid peptides (Aβ42), which aggregate and eventually paralyze the worm, significantly reduces toxicity [83]. Very interestingly, anti-PAK1 compounds such as caffeic acid phenethyl ester (CAPE) and artepillin C (ARC) in propolis boost the production of heat-shock proteins (HSP16 in this worm), and delay the onset of beta-amyloid-induced paralysis in the AD worm model [72], raising the possibility that propolis could be potentially useful for therapy of AD and/or HD. Caffeic acid and CAPE are also coffee components that may contribute to coffee's apparent neuroprotective activity, which has been suggested by the epidemiology of AD and PD.

Several years ago, signal transducers required for the progression of HD began to be unveiled. HDAC was among the first identified [84,85]. In 2003, a group at London University Hospital found that SAHA, an HDAC inhibitor used to treat a rare cancer called CTCL, delayed the onset of HD symptoms in a mouse model [84], suggesting that HDAC or one of its downstream signal transducers is essential for HD. In 2006, using an HD model of *C. elegans*, Bates and collaborators [85] (Massachusetts General Hospital, Boston) supported a role for HDAC. Thus, far more potent HDAC inhibitors such as FK228 could be potentially useful for the treatment of HD in the near future. FK228 suppresses polyQ-induced apoptosis in cell culture [86]. As discussed in Chapter 3, FK228 eventually blocks the oncogenic PAK1, suggesting the possibility that PAK1, downstream of HDAC, might be involved in HD. This is consistent with data showing that polyQ aggregation is facilitated by PAK1 [76] and that PIX, a direct activator of PAK1, also enhances polyQ aggregation, the cause of HD [77]. Independently, Apostol and collaborators [87] at the University of California (Davis) demonstrated that CEP-1347, a direct inhibitor of PAK1, limited HD symptoms in a mouse model. Thus, it is now clear that natural compounds that directly or indirectly limit protein aggregation and aberrant PAK1 activation such as trehalose, propolis, and curcumin support their testing in HD trials.

In summary, the PAK regulation of actin dynamics is dysregulated in several neurodegenerative diseases, including AD and HD. Because of the normal role of PAK in many cellular processes, an optimum therapeutic needs to correct aberrant PAK activity without preventing its normal function. Although highly selective and potent inhibitors should be tested in models, natural products with moderate inhibitory activity may be a safer choice for trials.

Acknowledgments

The authors are very grateful to Dr. Hiroshi Maruta for his critical reading of this manuscript and valuable comments. A part of this manuscript was published in the 2012 PAK issue of *Cellular Logistics* (edited by Jeff Field).

References

[1] Evans DA, Funkenstein HH, Albert MS, Scherr PA, Cook NR, Chown MJ, et al. Prevalence of Alzheimer's disease in a community population of older persons. Higher than previously reported. JAMA 1989;262:2551–6.

[2] Selkoe DJ. Alzheimer's disease is a synaptic failure. Science 2002;298(5594):789–91.

[3] Coleman P, Federoff H, Kurlan R. A focus on the synapse for neuroprotection in Alzheimer disease and other dementias. Neurology 2004;63(7):1155–62.

[4] Cleary JP, Walsh DM, Hofmeister JJ, Shankar GM, Kuskowski MA, Selkoe DJ, et al. Natural oligomers of the amyloid-beta protein specifically disrupt cognitive function. Nat Neurosci 2005;8(1):79–84.

[5] Lesne S, Koh MT, Kotilinek L, Kayed R, Glabe CC, Yang A, et al. A specific amyloid-β assembly in the brain impairs memory. Nature 2006;440(7082):352–7.

[6] Lue LF, Kuo YM, Roher AE, Brachova L, Shen Y, Sue L, et al. Soluble amyloid beta peptide concentration as a predictor of synaptic change in Alzheimer's disease. Am J Pathol 1999;155:853–62.

[7] McLean CA, Cherny RA, Fraser FW, Fuller SJ, Smith MJ, Beyreuther K, et al. Soluble pool of Aβ amyloid as a determinant of severity of neurodegeneration in Alzheimer's disease. Ann Neurol 1999;46:860–6.

[8] Klyubin I, Walsh DM, Lemere CA, Cullen WK, Shankar GM, Betts V, et al. Amyloid beta protein immunotherapy neutralizes Abeta oligomers that disrupt synaptic plasticity in vivo. Nat Med 2005;11(5):556–61.

[9] Ma QL, Yang F, Calon F, Ubeda OJ, Hansen JE, Weisbart RH, et al. p21-activated kinase-aberrant activation and translocation in Alzheimer disease pathogenesis. J Biol Chem 2008;283(20):14132–14143.

[10] Carlisle HJ, Kennedy MB. Spine architecture and synaptic plasticity. Trends Neurosci 2005;28(4):182–7.

[11] Schubert V, Dotti CG. Transmitting on actin: synaptic control of dendritic architecture. J Cell Sci 2007;120(Pt 2):205–12.

[12] Sekino Y, Kojima N, Shirao T. Role of actin cytoskeleton in dendritic spine morphogenesis. Neurochem Int 2007;51(2–4):92–104.

[13] Narumiya S, Ishizaki T, Watanabe N. Rho effectors and reorganization of actin cytoskeleton. FEBS Lett 1997;410(1):68–72.

[14] Hall A. Rho GTPases and the actin cytoskeleton. Science 1998;279(5350):509–14.

[15] Kirchhausen T, Rosen FS. Disease mechanism: unravelling Wiskott–Aldrich syndrome. Curr Biol 1996;6(6):676–8.

[16] Tang BL, Liou YC. Novel modulators of amyloid-beta precursor protein processing. J Neurochem 2007;100(2):314–23.

[17] Zhao L, Ma QL, Calon F, Harris-White ME, Yang F, Lim GP, et al. Role of p21-activated kinase pathway defects in the cognitive deficits of Alzheimer disease. Nat Neurosci 2006;9(2):234–42.

[18] Ramakers GJ. Rho proteins, mental retardation and the cellular basis of cognition. Trends Neurosci 2002;25(4):191–9.

[19] Bienvenu T, des Portes V, McDonell N, Carrie A, Zemni R, Couvert P, et al. Missense mutation in PAK3, R67C, causes X-linked nonspecific mental retardation. Am J Med Genet 2000;93(4):294–8.

[20] Allen KM, Gleeson JG, Bagrodia S, Partington MW, MacMillan JC, Cerione RA, et al. PAK3 mutation in nonsyndromic X-linked mental retardation. Nat Genet 1998;20(1):25–30.

[21] Hayashi ML, Choi SY, Rao BS, Jung HY, Lee HK, Zhang D, et al. Altered cortical synaptic morphology and impaired memory consolidation in forebrain-specific dominant-negative PAK transgenic mice. Neuron 2004;42(5):773–87.

[22] Meng Y, Zhang Y, Tregoubov V, Janus C, Cruz L, Jackson M, et al. Abnormal spine morphology and enhanced LTP in LIMK-1 knockout mice. Neuron 2002;35(1):121–33.

[23] Fiala JC, Spacek J, Harris KM. Dendriticd spine pathology: cause or consequence of neurological disorders?. Brain Res Brain Res Rev 2002;39(1):29–54.

[24] Harigaya Y, Shoji M, Shirao T, Hirai S. Disappearance of actin-binding protein, drebrin, from hippocampal synapses in Alzheimer's disease. J Neurosci Res 1996;43(1):87–92.

[25] Hatanpaa K, Isaacs KR, Shirao T, Brady DR, Rapoport SI. Loss of proteins regulating synaptic plasticity in normal aging of the human brain and in Alzheimer disease. J Neuropathol Exp Neurol 1999;58(6):637–43.

[26] Takahashi H, Sekino Y, Tanaka S, Mizui T, Kishi S, Shirao T. Drebrin-dependent actin clustering in dendritic filopodia governs synaptic targeting of postsynaptic density-95 and dendritic spine morphogenesis. J Neurosci 2003;23(16):6586–95.

[27] Biou V, Brinkhaus H, Malenka RC, Matus A. Interactions between drebrin and Ras regulate dendritic spine plasticity. Eur J Neurosci 2008;27(11):2847–59.

[28] Nguyen TV, Galvan V, Huang W, Banwait S, Tang H, Zhang J, et al. Signal transduction in Alzheimer disease: p21-activated kinase signaling requires C-terminal cleavage of APP at Asp664. J Neurochem 2008;104(4):1065–80.

[29] Arsenault D, Julien C, Tremblay C, Calon F. DHA improves cognition and prevents dysfunction of entorhinal cortex neurons in 3xTg-AD mice. PLoS One 2011;6(2):e17397.

[30] Mendoza-Naranjo A, Gonzalez-Billault C, Maccioni RB. Abeta1–42 stimulates actin polymerization in hippocampal neurons through Rac1 and Cdc42 Rho GTPases. J Cell Sci 2007;120(Pt 2):279–88.

[31] Mendoza-Naranjo A, Contreras-Vallejos E, Henriquez DR, Otth C, Bamburg JR, Maccioni RB, et al. Fibrillar amyloid-beta1–42 modifies actin organization affecting the cofilin phosphorylation state: a role for Rac1/cdc42 effector proteins and the slingshot phosphatase. J Alzheimers Dis 2012;29(1):63–77.

[32] Edwards DC, Sanders LC, Bokoch GM, Gill GN. Activation of LIM-kinase by Pak1 couples Rac/Cdc42 GTPase signalling to actin cytoskeletal dynamics. Nat Cell Biol 1999;1(5):253–9.

[33] Salminen A, Suuronen T, Kaarniranta K. ROCK, PAK, and Toll of synapses in Alzheimer's disease. Biochem Biophys Res Commun 2008;371(4):587–90.

[34] Bamburg JR, Wiggan OP. ADF/cofilin and actin dynamics in disease. Trends Cell Biol 2002;12(12):598–605.

[35] Heredia L, Helguera P, de Olmos S, Kedikian G, Sola Vigo F, LaFerla F, et al. Phosphorylation of actin-depolymerizing factor/cofilin by LIM-kinase mediates amyloid beta-induced degeneration: a potential mechanism of neuronal dystrophy in Alzheimer's disease. J Neurosci 2006;26(24):6533–42.

[36] Zhang Z, Ottens AK, Larner SF, Kobeissy FH, Williams ML, Hayes RL, et al. Direct Rho-associated kinase inhibition induces cofilin dephosphorylation and neurite outgrowth in PC-12 cells. Cell Mol Biol Lett 2006;11:12–29.

[37] Bliss TV, Collingridge GL. A synaptic model of memory: long-term potentiation in the hippocampus. Nature 1993;361(6407):31–9.

[38] Saneyoshi T, Wayman G, Fortin D, Davare M, Hoshi N, Nozaki N, et al. Activity-dependent synaptogenesis: regulation by a CaM-kinase kinase/CaM-kinase I/betaPIX signaling complex. Neuron 2008;57(1):94–107.

[39] Bi H, Sze CI. N-methyl-D-aspartate receptor subunit NR2A and NR2B messenger RNA levels are altered in the hippocampus and entorhinal cortex in Alzheimer's disease. J Neurol Sci 2002;200(1–2):11–18.

[40] Dewachter I, Filipkowski RK, Priller C, Ris L, Neyton J, Croes S, et al. Deregulation of NMDA-receptor function and downstream signaling in APP[V717I] transgenic mice. Neurobiol Aging 2009;30(2):241–56.

[41] Shankar GM, Bloodgood BL, Townsend M, Walsh DM, Selkoe DJ, Sabatini BL. Natural oligomers of the Alzheimer amyloid-beta protein induce reversible synapse loss by modulating an NMDA-type glutamate receptor-dependent signaling pathway. J Neurosci 2007;27(11):2866–75.

[42] De Felice FG, Wu D, Lambert MP, Fernandez SJ, Velasco PT, Lacor PN, et al. Alzheimer's disease-type neuronal tau hyperphosphorylation induced by A beta oligomers. Neurobiol Aging 2008;29(9):1334–47.

[43] Nakazawa T, Komai S, Tezuka T, Hisatsune C, Umemori H, Semba K, et al. Characterization of Fyn-mediated tyrosine phosphorylation sites on GluR epsilon 2 (NR2B) subunit of the N-methyl-D-aspartate receptor. J Biol Chem 2001;276(1):693–9.

[44] He H, Hirokawa Y, Levitzki A, Maruta H. An anti-Ras cancer potential of PP1, an inhibitor specific for Src family kinases: in vitro and in vivo studies. Cancer J. 2000;6(4):243–8.

[45] Schindler T, Sicheri F, Pico A, Gazit A, Levitzki A, Kuriyan J. Crystal structure of Hck in complex with a Src family-selective tyrosine kinase inhibitor. Mol Cell 1999;3(5):639–48.

[46] Yadav V, Denning MF. Fyn is induced by Ras/PI3K/Akt signaling and is required for enhanced invasion/migration. Mol Carcinog 2011;50(5):346–52.

[47] Park H, Ishihara D, Cox D. Regulation of tyrosine phosphorylation in macrophage phagocytosis and chemotaxis. Arch Biochem Biophys 2011;510(2):101–11.

[48] Neniskyte U, Neher JJ, Brown GC. Neuronal death induced by nanomolar amyloid β is mediated by primary phagocytosis of neurons by microglia. J Biol Chem 2011;286(46):39904–39913.

[49] He H, Hirokawa Y, Gazit A, Yamashita Y, Mano H, Kawakami Y, et al. The Tyr-kinase inhibitor AG879, that blocks the ETK-PAK1 interaction, suppresses the RAS-induced PAK1 activation and malignant transformation. Cancer Biol Ther 2004;3(1):96–101.

[50] Hirokawa Y, Levitzki A, Lessene G, Baell J, Xiao Y, Zhu H, et al. Signal therapy of human pancreatic cancer and NF1-deficient breast cancer xenograft in mice by a combination of PP1 and GL-2003, anti-PAK1 drugs (Tyr-kinase inhibitors). Cancer Lett 2007;245(1–2):242–51.

[51] He H, Hirokawa Y, Manser E, Lim L, Levitzki A, Maruta H. Signal therapy for RAS-induced cancers in combination of AG 879 and PP1, specific inhibitors for ErbB2 and Src family kinases, that block PAK activation. Cancer J. 2001;7(3):191–202.

[52] Jones RL, Judson IR. The development and application of imatinib. Expert Opin Drug Saf 2005;4(2):183–91.

[53] Nheu TV, He H, Hirokawa Y, Tamaki K, Florin L, Schmitz ML, et al. The K252a derivatives, inhibitors for the PAK/MLK kinase family selectively block the growth of RAS transformants. Cancer J 2002;8(4):328–36.

[54] Murray BW, Guo C, Piraino J, Westwick JK, Zhang C, Lamerdin J, et al. Small-molecule p21-activated kinase inhibitor PF-3758309 is a potent inhibitor of oncogenic signaling and tumor growth. Proc Natl Acad Sci USA 2010;107(20):9446–51.

[55] Bozyczko-Coyne D, O'Kane TM, Wu ZL, Dobrzanski P, Murthy S, Vaught JL, et al. CEP-1347/KT-7515, an inhibitor of SAPK/JNK pathway activation, promotes survival and blocks multiple events associated with Abeta-induced cortical neuron apoptosis. J Neurochem 2001;77(3):849–63.

[56] Yang F, Lim GP, Begum AN, Ubeda OJ, Simmons MR, Ambegaokar SS, et al. Curcumin inhibits formation of amyloid beta oligomers and fibrils, binds plaques, and reduces amyloid in vivo. J Biol Chem 2005;280(7):5892–901.

[57] Lim GP, Chu T, Yang F, Beech W, Frautschy SA, Cole GM. The curry spice curcumin reduces oxidative damage and amyloid pathology in an Alzheimer transgenic mouse. J Neurosci 2001;21(21):8370–7.

[58] Gota VS, Maru GB, Soni TG, Gandhi TR, Kochar N, Agarwal MG. Safety and pharmacokinetics of a solid lipid curcumin particle formulation in osteosarcoma patients and healthy volunteers. J Agric Food Chem 2010;58(4):2095–9.

[59] Jing F, Mogi M, Sakata A, Iwanami J, Tsukuda K, Ohshima K, et al. Direct stimulation of angiotensin II type 2 receptor enhances spatial memory. J Cereb Blood Flow Metab 2012;32(2):248–55.

[60] Wan Y, Wallinder C, Plouffe B, Beaudry H, Mahalingam AK, Wu X, et al. Design, synthesis, and biological evaluation of the first selective nonpeptide AT2 receptor agonist. J Med Chem 2004;47(24):5995–6008.

[61] Steckelings UM, Larhed M, Hallberg A, Widdop RE, Jones ES, Wallinder C, et al. Nonpeptide AT2-receptor agonists. Curr Opin Pharmacol 2011;11(2):187–92.

[62] Kim JS, Huang TY, Bokoch GM. Reactive oxygen species regulate a slingshot-cofilin activation pathway. Mol Biol Cell 2009;20(11):2650–60.

[63] Zhu F, Qian C. Berberine chloride can ameliorate the spatial memory impairment and increase the expression of interleukin-1beta and inducible nitric oxide synthase in the rat model of Alzheimer's disease. BMC Neurosci 2006;7:78. doi: 1471-2202-7-78

[64] Bahar M, Deng Y, Zhu X, He S, Pandharkar T, Drew ME, et al. Potent antiprotozoal activity of a novel semi-synthetic berberine derivative. Bioorg Med Chem Lett 2011;21(9):2606–10.

[65] Cramer PE, Cirrito JR, Wesson DW, Lee CY, Karlo JC, Zinn AE, et al. ApoE-directed therapeutics rapidly clear beta-amyloid and reverse deficits in AD mouse models. Science 2012;335(6075):1503–6.

[66] Bai X, Cerimele F, Ushio-Fukai M, Waqas M, Campbell PM, Govindarajan B, et al. Honokiol, a small molecular weight natural product, inhibits angiogenesis in vitro and tumor growth in vivo. J Biol Chem 2003;278(37):35501–35507.

[67] Kotani H, Tanabe H, Mizukami H, Amagaya S, Inoue M. A naturally occurring rexinoid, honokiol, can serve as a regulator of various retinoid x receptor heterodimers. Biol Pharm Bull 2012;35(1):1–9.

[68] Rubinsztein DC, Leggo J, Coles R, Almqvist E, Biancalana V, Cassiman JJ, et al. Phenotypic characterization of individuals with 30–40 CAG repeats in the Huntington disease (HD) gene reveals HD cases with 36 repeats and apparently normal elderly individuals with 36–39 repeats. Am J Hum Genet 1996;59(1):16–22.

[69] Ho LW, Brown R, Maxwell M, Wyttenbach A, Rubinsztein DC. Wild type Huntingtin reduces the cellular toxicity of mutant Huntingtin in mammalian cell models of Huntington's disease. J Med Genet 2001;38(7):450–2.

[70] Leavitt BR, Guttman JA, Hodgson JG, Kimel GH, Singaraja R, Vogl AW, et al. Wild-type huntingtin reduces the cellular toxicity of mutant huntingtin in vivo. Am J Hum Genet 2001;68(2):313–24.

[71] Rigamonti D, Sipione S, Goffredo D, Zuccato C, Fossale E, Cattaneo E. Huntingtin's neuroprotective activity occurs via inhibition of procaspase-9 processing. J Biol Chem 2001;276(18):14545–14548.

[72] Rigamonti D, Bauer JH, De-Fraja C, Conti L, Sipione S, Sciorati C, et al. Wild-type huntingtin protects from apoptosis upstream of caspase-3. J Neurosci 2000;20(10):3705–13.

[73] Luo S, Rubinsztein DC. Huntingtin promotes cell survival by preventing Pak2 cleavage. J Cell Sci 2009;122(Pt 6):875–85.

[74] Cotteret S, Jaffer ZM, Beeser A, Chernoff J. p21-Activated kinase 5 (Pak5) localizes to mitochondria and inhibits apoptosis by phosphorylating BAD. Mol Cell Biol 2003;23(16):5526–39.

[75] Nagai Y, Inui T, Popiel HA, Fujikake N, Hasegawa K, Urade Y, et al. A toxic monomeric conformer of the polyglutamine protein. Nat Struct Mol Biol 2007;14(4):332–40.

[76] Luo S, Mizuta H, Rubinsztein DC. p21-activated kinase 1 promotes soluble mutant huntingtin self-interaction and enhances toxicity. Hum Mol Genet 2008;17(6):895–905.

[77] Eriguchi M, Mizuta H, Luo S, Kuroda Y, Hara H, Rubinsztein DC. alpha Pix enhances mutant huntingtin aggregation. J Neurol Sci 2010;290(1–2):80–5.

[78] Tanaka M, Machida Y, Niu S, Ikeda T, Jana NR, Doi H, et al. Trehalose alleviates polyglutamine-mediated pathology in a mouse model of Huntington disease. Nat Med 2004;10(2):148–54.

[79] Sarkar S, Davies JE, Huang Z, Tunnacliffe A, Rubinsztein DC. Trehalose, a novel mTOR-independent autophagy enhancer, accelerates the clearance of mutant huntingtin and alpha-synuclein. J Biol Chem 2007;282(8):5641–52.

[80] Sarkar S, Rubinsztein DC. Small molecule enhancers of autophagy for neurodegenerative diseases. Mol Biosyst 2008;4(9):895–901.

[81] Arguelles JC. Thermotolerance and trehalose accumulation induced by heat shock in yeast cells of Candida albicans. FEMS Microbiol Lett 1997;146(1):65–71.

[82] Honda Y, Tanaka M, Honda S. Trehalose extends longevity in the nematode Caenorhabditis elegans. Aging Cell 2010;9(4):558–69.

[83] Fonte V, Kipp DR, Yerg III J, Merin D, Forrestal M, Wagner E, et al. Suppression of in vivo beta-amyloid peptide toxicity by overexpression of the HSP-16.2 small chaperone protein. J Biol Chem 2008;283(2):784–91.

[84] Hockly E, Richon VM, Woodman B, Smith DL, Zhou X, Rosa E, et al. Suberoylanilide hydroxamic acid, a histone deacetylase inhibitor, ameliorates motor deficits in a mouse model of Huntington's disease. Proc Natl Acad Sci USA 2003;100(4):2041–6.

[85] Bates EA, Victor M, Jones AK, Shi Y, Hart AC. Differential contributions of Caenorhabditis elegans histone deacetylases to huntingtin polyglutamine toxicity. J Neurosci 2006;26(10):2830–8.

[86] Shimohata M, Shimohata T, Igarashi S, Naruse S, Tsuji S. Interference of CREB-dependent transcriptional activation by expanded polyglutamine stretches—augmentation of transcriptional activation as a potential therapeutic strategy for polyglutamine diseases. J Neurochem 2005;93(3):654–63.

[87] Apostol BL, Simmons DA, Zuccato C, Illes K, Pallos J, Casale M, et al. CEP-1347 reduces mutant huntingtin-associated neurotoxicity and restores BDNF levels in R6/2 mice. Mol Cell Neurosci 2008;39(1):8–20.

7 PAK1 Controls the Lifespan

Sumino Yanase[1], and Hiroshi Maruta[2]

[1]Daito Bunka University, Saitama, Japan, [2]NF/TSC Cure Organisation, Melbourne, Australia

Abbreviations

AD	Alzheimer's disease
AMPK	AMP-activated kinase
CA	caffeic acid
CAPE	caffeic acid phenethyl ester
CR	calorie restriction
DN	dominant negative
HD	Huntington's disease
IR	insulin-like receptor
NF	neurofibromatosis
3-NPA	3-nitropropionic acid
RA	rosmarinic acid
TSC	tuberous sclerosis complex

7.1 Introduction

Calorie restriction (CR) significantly extends the lifespan of mice by activating the tumor-suppressing AMP-activated kinase (AMPK) through another tumor-suppressing kinase, LKB1 [1,2].

Very interestingly, LKB1 inactivates the oncogenic kinase PAK1 [3]. AMPK is known to activate the tumor-suppressing transcription factor FOXO, which is essential for longevity [4], whereas PAK1 inactivates FOXO [5]. Thus, it is most likely that CR extends the lifespan by activating FOXO through two distinct pathways involving AMPK and PAK1, which are the common targets of LKB1. In 2002, a Korean group reported that the aging process in mice strongly enhanced all three mitogen-activated protein kinases (MAPKs), namely extracellular signal-related kinases (ERK), c-Jun N-terminal kinases (JNK), and p38 [6], all of which are targets of PAK1. In contrast, CR markedly suppresses the age-related activation of these MAPKs [6], strongly suggesting, if not proving as yet, that CR inactivates PAK1.

PAKs, RAC/CDC42 (p21)-activated Kinases. DOI: http://dx.doi.org/10.1016/B978-0-12-407198-8.00007-2

In a tiny nematode called *Caenorhabditis elegans*, a strain called AGE-1, which is PI-3 kinase-deficient, can live almost twice as long as the control N2 [7]. Furthermore, an AKT-deficient mutant will live around 50% longer than the control N2 [8]. Both PI-3 kinase and AKT shorten the lifespan by inactivating FOXO [8]. Since PI-3 kinase activates both AKT and PAK1, it is conceivable that the difference in apparent lifespan between PI-3 kinase deficiency and AKT deficiency is attributable to the PAK1 deficiency.

Here we shall discuss the potential contribution of PAK1 to the lifespan of both mammals such as mice, and invertebrates such as *C. elegans* and *Drosophila*.

7.2   Snell Dwarf Mice

Mutation of several distinct genes is known to be responsible for the longevity of mice [9]. One of these genes is called Pit-1, and its dysfunction causes a long-living dwarf called Snell that weighs only one-third as much as normal mice but can live up to 30–40% longer than its normal counterparts, extending the lifespan from an average of about 900 days to an average of about 1200 days [9]. This dwarf mouse is resistant to a variety of diseases and oxidative stresses such as 3-nitropropionic acid (3-NPA). In normal mice, 3-NPA activates MEK/ERK kinases and stimulates a robust phosphorylation of JNK at Ser63, whereas no phosphorylation takes place in Snell mice, in response to 3-NPA [10]. Since both MEK/ERK and JNK are direct targets of PAK1, it is almost certain that in Snell mice, PAK1 is somehow blocked.

7.3   RAC-Deficient *Drosophila*

The GTPase RAC is the major activator of PAK1. Thus, when the dominant negative (DN) mutant of RAC is expressed in adult fruit flies (*Drosophila*), PAK1 is blocked in this transgenic fly. According to Yi Zhong's group [11] at Cold Spring Harbor Laboratories (CSHL), this RAC-deficient fly is extremely resistant to a variety of stresses such as oxidants, desiccation, starvation, and heat. In fact, this transgenic fly, in which PAK1 is blocked, can live for 40 days at 30°C, almost twice as long as the control fly (which lives for only 22 days).

7.4   PAK1 Blockers Extend the Lifespan of Tiny Animals

As was discussed in Chapter 3, curcumin is among the natural PAK1 blockers [12] that are available on the market rather inexpensively, although it has never been clinically used because of its poor bioavailability. However, it works *in vivo* in tiny animals such as fruit flies (*Drosophila*) and nematodes (*C. elegans*) to extend their lifespan. In 2010, a Korean group provided the first evidence that curcumin can extend the lifespan of the fruit fly *Drosophila* [13]. More recently a group at

National Taiwan University also reported that curcumin significantly extends the lifespan of the nematode *C. elegans* [14]. These findings strongly suggest, if not proving as yet, the possibility that PAK1 shortens the lifespan of these tiny animals, at least. Furthermore, two other natural PAK1 blockers, salidroside and caffeic acid (CA), as well as a CA dimer called rosmarinic acid (RA), were also shown to extend significantly the lifespan of *C. elegans* by activating the transcription factor FOXO [15,16]. Thus, it is most likely that caffeic acid phenethyl ester (CAPE), a natural CA derivative found in propolis, which also inactivates PAK1, could extend the lifespan of this worm.

7.4.1 C. elegans Genes Involved in Longevity

Since *C. elegans* is among the shortest lived experimental animals, and its whole genome has been sequenced, this tiny animal has been among the most frequently used invertebrates for studying lifespan-controlling genes. During the last two decades, several distinct genes that control lifespan have been identified in this nematode. The first gene is called AGE-1; it encodes the catalytic subunit p110 of PI-3 kinase, which phosphorylates a phospholipid called PIP2, producing an oncogenic PIP3 [7]. Dysfunction of AGE-1 extends the average lifespan of this worm by 40–65%, and its maximum life by 60–110%, depending on the temperature (20–25°C). In other words, AGE-1 (an oncogenic kinase) shortens the lifespan of this worm. Thus, in theory, any PI-3 kinase inhibitor, such as myricetin, could effectively extend the lifespan of this worm. However, it was shown recently by a German group in Munich that myricetin (even at 100 μM) only extends the average lifespan of this worm by 15% [17]. The IC_{50} of myricetin for mammalian PI-3 kinase is around 1.5 μM. Thus, we wonder whether the rather marginal lifespan-extending effect of this compound in the worm is due to its potentially much higher IC_{50} for *C. elegans* PI-3 kinase (AGE-1), or whether, instead, this compound has unknown side effects on this worm at this dose (100 μM). A second gene that shortens the lifespan of this worm is insulin-like receptor (IR), a cell surface Tyr kinase, whose dysfunction extends the lifespan by almost 100% [18]. Do these two genes share any common signal transducer that controls lifespan? Yes: IR activates AGE-1, which in turn inactivates the tumor-suppressive transcription factor FOXO (DAF-16) [8,18,19]. Dysfunction of the DAF-18 gene encoding PTEN [20], a tumor-suppressive PIP3 phosphatase, shortens the lifespan of this worm without causing development of any detectable tumors, indicating that like FOXO, PTEN is a life-extending signal transducer that might be called an elixir [19]. A natural aphrodisiac flavonol called icariin or its metabolite (icariside II), which, like thalidomide, upregulates PTEN and eventually activates FOXO, that could extend significantly the healthy lifespan of *C. elegans* [21,22].

AGE-1 activates two oncogenic kinases, AKT and PAK1. AKT is known to be essential for infection by pathogens [8]. Thus, dysfunction of AKT would contribute to the resistance of this worm to deadly pathogens, leading to a longer lifespan. So far no direct evidence has been shown for the contribution of PAK1 to the lifespan of this worm. However, some indirect evidence suggests that PAK1 could shorten the lifespan of this worm. First of all, in mammals PAK1 inactivates FOXO, which is essential for longevity [5]. There are more than two dozen natural compounds that block PAK1 and activate another kinase called AMPK. AMPK is a tumor-suppressive kinase that activates FOXO, leading to the extension of lifespan [4]. In fact, some of these natural PAK1 blockers (=AMPK activator) such as curcumin, CA, and salidroside were shown to extend the lifespan of this worm, as has been discussed earlier in this chapter.

7.4.2   Does PAK1 Shorten the Lifespan of C. elegans?

Back in 2007, we found that natural PAK1 blockers derived from propolis, such as CAPE and ARC, activate a heat-shock gene called *HSP*16.2 in *C. elegans*, and make this worm heat resistant [23], clearly indicating that PAK1 inactivates this heat-shock gene. Since the expression of this gene requires FOXO in this worm [24] and at least in mammals, PAK1 inactivates FOXO [5], it is conceivable that PAK1 suppresses the *HSP*16.2 gene by inactivating FOXO in this worm as well. Furthermore, it is well known that the activation of FOXO-*HSP*16.2 signaling leads to the longevity of this worm [24–26]. Thus, around the end of 2011, we started examining whether a PAK1-deficient strain of *C. elegans* called RB689 can live significantly longer than the wild-type N2, and whether the transcription factor FOXO is more nuclear localized (activated) in RB689 than in N2.

Quite interestingly, the litter size (the number of eggs laid) or self-fertility of RB689 is markedly small, only around 15% of that of N2 [23]. This phenotype is very similar to that of CAPE/ARC-treated N2 and the nontreated AGE-1-deficient mutant, whose fertility is only around 25% of the control N2 [7,23]. Furthermore, it was shown more than two decades ago that the AGE-1-deficient mutant of this worm is much more thermotolerant than N2 at 30–35°C [24], just like RB689 (PAK1-deficient) and CAPE/ARC-treated N2 [23]. These observations clearly indicate that both the litter size and thermotolerance of this worm are controlled by PI-3 kinase, mainly through PAK1, because so far there is no evidence for AKT to control either litter size or thermotolerance of this worm.

7.4.3   How About Lifespan and FOXO?

Finally, in the beginning of 2012, we managed to demonstrate that the mean lifespan of RB689 (around 27 days) is clearly much longer than that of N2 (17–22 days) by 20–60% (for details, see Figure 7.1), confirming that PAK1 indeed shortens the lifespan of this worm (Yanase, S. et al., unpublished observation). Furthermore, the nuclear localization of FOXO in RB689 is significantly greater than that in N2 (data not shown), confirming that PAK1 inactivates FOXO in this worm, as it does in mammals [5]. Does PAK1 shorten the lifespan of mammals, too? Around the end of 2011 we heard a rumor that a group at the National Institute on Aging (NIA) in Baltimore has started measuring the lifespan of a PAK1-deficient strain of mice [27] compared with that of control counterparts. Since the nematode and mouse share basically the same signaling pathways around PAK1 in all organs except for the cardiovascular system, which the nematode clearly lacks, we would expect a similar outcome as long as PAK1-null mice do not confront any stress that causes cardiac

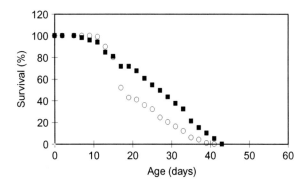

Figure 7.1 The effect of PAK1 deficiency on the lifespan. The survival rate between N2 (the control) (circle) and RB689 (PAK1-deficient) (square) strains of *C. elegans* was compared at 20°C on the standard solid plate. Fifty percent of RB689 survived to day 27, 10 days longer than the N2 counterparts.

hypertrophy. However, the average lifespan of normal mice is two-three years (around 900 days), which is around 50 times as long as that of this worm, and therefore if the PAK1 deficiency extends their lifespan, just like the dysfunction of Pit-1 gene does in Snell dwarf mice [9], this mouse lifespan study would take at least four-five years (around 1600 days).

7.5   Crosstalk Between PI-3 kinase-PAK1 and TOR-S6K Signaling Pathways

In this context, it should be worth noting that TOR-deficient mutants of *Drosophila* and *C. elegans* also clearly live longer than their control counterparts [28,29]. Furthermore, rapamycin, a TOR inhibitor, extends the lifespan of both *Drosophila* and mice [29,30]. Just like AGE-1 (PI-3 kinase)-deficient and PAK1-deficient mutants of *C. elegans*, the TOR-deficient nematode also shows a marked reduction in the litter size and increased thermotolerance, and all of these effects depend on FOXO [28]. Thus, it is quite conceivable that the PI-3 kinase-PAK1 signaling pathway is closely linked to the TOR signaling pathway in controlling lifespan, fertility, and thermotolerance. Interestingly, rapamycin's life-extending effect in *Drosophila* is abolished by the constitutively active mutant of S6K, the major substrate/effector of the kinase TOR [28], clearly indicating that the TOR-S6K signaling pathway is responsible for shortening lifespan. Furthermore, it was revealed recently that the kinase S6K activates PAK1 [31], and its downstream kinase MEK activates the kinase S6K [32], which in turn activates MPKs (MAP kinases) such as ERK, p38, and JNK downstream of the PAK1-MEK signaling. Thus, S6K, PAK1, and MEK form a vicious oncogenic cycle (Figure 7.2). Furthermore, ERK activates TOR by inactivating the tumor suppressor *TSC*2 [33], an RheB GAP, which normally blocks TOR [34], forming another vicious oncogenic cycle.

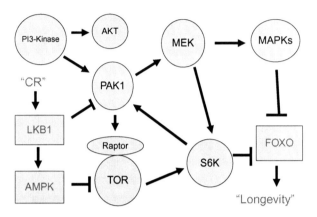

Figure 7.2 Crosstalk between PAK1 and TOR pathways.

Thus, it is most likely that these MAPKs are the common effectors of both onco-genic TOR-S6K and PI-3 kinase-PAK1-MEK-ERK pathways (Figure 7.2), which play the main role in both promoting tumor growth and shortening lifespan. In fact, at least ERK and p38 were shown to phosphorylate FOXO directly for its inactiva-tion [35]. In addition, PAK1 appears to control the interaction of TOR with a third protein called raptor, which is essential for full activation of TOR. Curcumin, a direct PAK1 inhibitor [12], was shown to disassemble the TOR-raptor complex [36]. Furthermore, FOXO downregulates raptor expression [37].

Back in 2005, David Gutmann's group [38] at Washington University (St. Louis, MO) revealed a rather surprising discovery: TOR is hyperactivated in *NF*1-deficient tumors, and rapamycin, which inhibits TOR, blocks the anchorage-independent growth of these tumor cells.

In other words, the growth of these tumors depends not only on PAK1 but also on TOR.

How does the dysfunction of the *NF*1 gene lead to the hyperactivation of TOR? As discussed in Chapters 3 and 5, the *NF*1 gene product is an attenuator of both RAS and RAC, and therefore in *NF*1-deficient tumors the oncogenic RAS-PI3 kinase-PAK1/AKT pathway is hyperactivated. Furthermore, it is known that the oncogenic kinase AKT inactivates the tumor suppresser *TSC*2 [39], and therefore like TSC tumors in which the TSC complex is inactivated, *NF*1-deficient tumors could harbor hyperactivated TOR (Figure 7.3). Alternatively, just as discussed above, PAK1 could hyperactivate TOR by inactivating FOXO, which normally downregulates raptor, essential for the full activation of TOR (Figure 7.2).

In other words, in principle either PAK1 blockers or TOR blockers would extend the lifespan as well as suppress the growth of cancers or NF/TSC tumors by inacti-vating both TOR-S6K and MEK-ERK signaling pathways. However, unlike PAK1 blockers, TOR inhibitors such as rapamycin and affinitor are immunosuppressors and also cause hypertension, and thus might not be suitable for the lifelong treatment of all cancer/NF/TSC patients or the life extension of apparently healthy people.

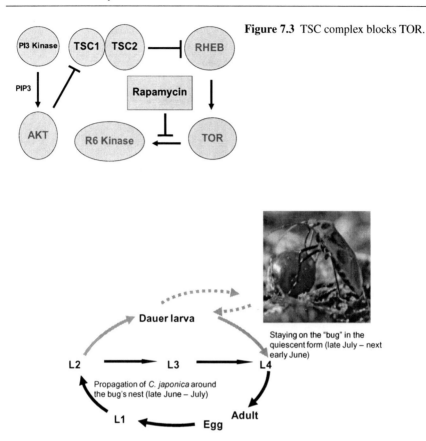

Figure 7.3 TSC complex blocks TOR.

Figure 7.4 Life cycle of *C. japonica*. Kindly provided by Dr. Ryusei Tanaka.

7.6 The Nematode *Caenorhabditis japonica* as a Quicker Animal Model for Elixir Screening

The dauer larvae (DL) of the nematode *C. japonica*, closely related to *C. elegans*, have been known to keep species-specific phoretic associations with the shield bug *Parastrachia japonensis* as shown in Figure 7.4 [40]. Recently a Japanese group led by Ryusei Tanaka at Saga University found that its DL have a much longer lifespan than their *C. elegans* counterparts in a biological setting (as long as the former being associated with this bug). However, once *C. japonica* DL were detached from their phoretic host, they did not survive for more than 10 days, while more than 50% of *C. elegans* survived for over 20 days under the same conditions [40]. It still remains to be clarified how this host provides *C. japonica* with its extremely long survival, and we are currently testing whether this bug secretes any substance that suppresses PAK1 or its further upstream activators such as RAC/CDC42 and PI-3 kinase. Nevertheless, it is quite conceivable that the detached (host-free) DL of *C. japonica*

Figure 7.5 Michael Rose.

will in the future serve as an ideal short-lived animal model for a very quick elixir screening.

7.7   Antiaging (Methuselah) Venture

Several years ago Michael Rose (Figure 7.5), working at University of California, Irvine, published an inspiring book entitled *The Long Tomorrow: How Advances in Evolutionary Biology Can Help Us Postpone Aging* [41]. He is a pioneer in aging research using *Drosophila*, a fruit fly, and around 1980 he created the first long-lived *Drosophila* strain, called Methuselah, which lives longer but has a significantly smaller litter size than its wild-type counterpart [42].

More than a decade later his group found that the Methuselah is significantly more resistant to a variety of stresses, such as starvation and desiccation, than its wild-type counterpart [43]. Interestingly, all these phenotypes of Methuselah are basically identical to those of RB689 (PAK1-KO) strain of *C. elegans*, suggesting the possibility that Methuselah might lack PAK1, or its upstream kinases such as PI-3 kinase, in part. Although Rose's group has been involved in the development of an effective elixir (antiaging drug) with several venture business groups, so far none of these projects has been successful. He speculates the major cause of their failure must be the lack of a so-called "killer application." For unlike a variety of cancers, aging *per se* does not cause any immediate threat to our life. However, if a variety of PAK1 blockers such as propolis and curcumin are proven in the future to suppress not only the growth of cancers but also the aging process without any side effects, the Methuselah venture will have its killer application, and ultimately should flourish.

Acknowledgments

One of the authors (HM) is very grateful to Drs. Yuan Luo of University of Maryland at Baltimore and Warwick Grant of La Trobe University in Melbourne for kindly providing him with laboratory space and support during his study of the PAK1-deficient (RB689) and *HSP*16.2-GFP transgenic (CL2070) strains of *C. elegans* in 2007.

Another author (SY) thanks Dr. Masamitsu Fukuyama of the University of Tokyo and the *C. elegans* Genome Center (CGC) for kindly providing us with RB689 and CL2070 strains, respectively.

References

[1] Canto C, Auwerx J. Calorie restriction: is AMPK a key sensor and effector?. Physiology 2011;26:214–24.

[2] Shaw R, Kosmatka S, Bardeesy N, Hurley R, et al. The tumor suppressor LKB1 kinase directly activates AMP-activated kinase and regulates apoptosis in response to energy stress. Proc Natl Acad Sci USA 2004;101:3329–35.

[3] Deguchi A, Miyashi H, Kojima Y, Okawa K, et al. LKB1 Suppresses p21-activated Kinase-1 (PAK1) by Phosphorylation of Thr109 in the p21-binding Domain. J Biol Chem 2010;285:18283–90.

[4] Greer E, Dowlatshahi D, Banko M, Villen J, et al. An AMPK-FOXO pathway mediates longevity induced by a novel method of dietary restriction in *C. elegans*. Curr Biol 2007;17:1646–56.

[5] Vadlamudi R, Kumar R. p21-activated kinase 1 (PAK1): an emerging therapeutic target. Cancer Treat Res 2004;119:77–88.

[6] Kim HJ, Jung KJ, Yu BP, Cho CG, et al. Influence of aging and calorie restriction on MAPKs activity in rat kidney. Exp Gerontol 2002;37:1041–53.

[7] Friedman D, Johnson T. A mutation in the Age-1 gene in *Caenorhabditis elegans* lengthens life and reduces hermaphrodite fertility. Genetics 1988;118:75–86.

[8] Gami M, Iser W, Hanselman K, Wolkow C. Activated AKT signaling in *C. elegans* uncouples temporally distinct outputs of DAF-2 signalling. BMC Dev Biol 2006;6:45.

[9] Vergara M, Smith-Wheelock M, Harper J, Sigler R, et al. Hormone-treated Snell dwarf mice regain fertility but remain long lived and disease resistant. J Gerontol A Biol Sci Med Sci 2004;59:1244–50.

[10] Madsena M, Hsieha CC, Boylstona W, Kevin Flurkeyb K, et al. Altered oxidative stress response of the long-lived Snell dwarf mouse. Biochem Biophys Res Commun 2004;318:998–1005.

[11] Shuai Y, Yisi Zhang Y, Gao L, Zhong. Y. Stress resistance conferred by neuronal expression of dominant-negative Rac in Adult *Drosophila melanogaster*. J Neurogenet 2011;25:35–9.

[12] Cai XZ, Wang J, Li XD, Wang GL, et al. Curcumin suppresses proliferation and invasion in human gastric cancer cells by down-regulation of PAK1 activity and cyclin D1 expression. Cancer Biol Ther 2009;8:1360–8.

[13] Lee KS, Lee BS, Semnani S, Avanesian A, et al. Curcumin extends life span, improves health span, and modulates the expression of age-associated aging genes in *Drosophila melanogaster*. Rejuven Res 2010;13:561–70.

[14] Liao V, Yu CW, Chu YJ, Li WH, et al. Curcumin-mediated lifespan extension in *Caenorhabditis elegans*. Mech Aging Dev 2011;132:480–7.

[15] Wiegant F, Surinova S, Ytsma E, Langelaar-Makkinje M, et al. Plant adaptogens increase lifespan and stress resistance in *C. elegans*. Biogerontology 2009;10:27–42.

[16] Pietsch K, Saul N, Shumon Chakrabarti S, Stürzenbaum S, et al. Hormetins, antioxidants and prooxidants: defining quercetin-, caffeic acid- and rosmarinic acid-mediated life extension in *C. elegans*. Biogerontology 2011;12:329–47.

[17] Gruenz G, Haas K, Soukup S, Klingenspor M, et al. Structural features and bioavail-ability of four flavonoids and their implications for lifespan-extending and antioxidant functions in *C. elegans*. Mech Aging Dev 2012;133:1–10.

[18] Kenyon C, Chang J, Gensch E, Rudner A, et al. A *C. elegans* mutant that lives twice as long as wild type. Nature 1993;366:461–4.

[19] Dorman J, Albinder B, Shroyer T, Kenyon C. The AGE-1 and DAF-2 genes function in a common pathway to control the lifespan of *Caenorhabditis elegans*. Genetics 1995;141:1399–406.

[20] Gil E, Link E, Liu L, Johnson C, et al. Regulation of the insulin-like developmental pathway of *Caenorhabditis elegans* by a homolog of the *PTEN* tumor suppressor gene. Proc Natl Acad Sci USA 1999;96:2925–30.

[21] Kim SH, Ahn KS, Jeong SJ, Kwon TR, et al. Janus activated kinase 2/signal transducer and activator of transcription 3 pathway mediates icariside II-induced apoptosis in U266 multiple myeloma cells. Eur J Pharmacol 2011;654:10–16.

[22] Cai WJ, Huang JH, Zhang SQ, Wu B, et al. Icariin and its derivative Icariside II extend healthspan via insulin/IGF-1 pathway in *C. elegans*. PLoS One 2011;6:e28835.

[23] Maruta H. An innovated approach to *in vivo* screening for the major anticancer drugs Horizons in cancer research, 41. Nova Science Publishers; 2010. 249–59

[24] Lithgow G, White T, Hinerfeld D, Johnson T. Thermotolerance of a long-lived mutant of *Caenorhabditis elegans*. J Gerontol 1994;49:B270–6.

[25] Walker G, Lithgow G. Lifespan extension in *C. elegans* by a molecular chaperone dependent upon insulin-like signals. Aging Cell 2003;2:131–9.

[26] Rea S, Wu D, Cyper J, Vaupel J, et al. A stress-sensitive reporter predicts longevity in isogenic populations of *Caenorhabditis elegans*. Nat Genet 2005;37:394–8.

[27] Allen J, Jaffer Z, Park S, Burgin S, et al. PAK1 regulates mast cell degranulation via effects on calcium mobilization and cytoskeletal dynamics. Blood 2009;113:2695–705.

[28] Hansen M, Taubert S, Crawford D, Libina N, et al. Lifespan extension by conditions that inhibit translation in *Caenorhabditis elegans*. Aging Cell 2007;6:95–110.

[29] Bjedov I, Toivonen J, Kerr F, Slack C, et al. Mechanisms of life span extension by rapa-mycin in the fruit fly *Drosophila melanogaster*. Cell Metab 2010;11:35–46.

[30] Anisimov V, Zabezhinski M, Popovich I, Piskunova T, et al. Rapamycin increases lifespan and inhibits spontaneous tumorigenesis in inbred female mice. Cell Cycle 2011;10:4230–6.

[31] Ip CK, Cheung AN, Ngan HY, Wong AS. p70 S6 kinase in the control of actin cytoskeleton dynamics and directed migration of ovarian cancer cells. Oncogene 2011;30:2420–32.

[32] Wang L, Gout I, Proud C. Cross-talk between the ERK and p70 S6 kinase (S6K) sign-aling pathways: MEK-dependent activation of S6K2 in cardiomyocytes. J Biol Chem 2001;276:32670–32677.

[33] Ma L, Chen Z, Erdjument-Bromage H, Tempst P, et al. Phosphorylation and functional inactivation of TSC2 by Erk: implications for tuberous sclerosis and cancer pathogen-esis. Cell 2005;121:179–93.

[34] Manning B, Cantley L. Rheb fills a GAP between TSC and TOR. Trends Biochem Sci 2003;28:573–6.

[35] Asada S, Hiroaki Daitoku H, Matsuzaki H, Saito T, et al. Mitogen-activated protein kinases, Erk and p38, phosphorylate and regulate Foxo1. Cell Signal 2007;19:519–27.

[36] Beevers C, Chen L, Liu L, Luo Y, et al. Curcumin disrupts the mammalian target of rapamycin (TOR)-raptor complex. Cancer Res 2009;69:1000–8.

[37] Jia K, Chen D, Riddle D. The TOR pathway interacts with the insulin signaling pathway to regulate *C. elegans* larval development, metabolism and life span. Development 2004;131:3897–906.

[38] Dasgupta B, Yi YJ, Chen D, Weber J, et al. Hyperactivation of the TOR pathway in NF1-associated human and mouse brain tumors. Cancer Res 2005;65:2755–60.

[39] Inoki K, Li Y, Zhu T, Wu J, et al. TSC2 is phosphorylated and inhibited by Akt and suppresses mTOR signalling. Nat Cell Biol 2002;4:648–57.

[40] Tanaka R, Okumura E, Kanzaki N, Yoshiga T. Low survivorship of dauer larva in the nematode *Caenorhabditis japonica*, a potential comparative system for a model organism, *C. elegans*. Exp Gerontol 2012;47:388–93.

[41] Rose M.. In: In the long tomorrow: how can advances in evolutionary biology help us postpone aging. Oxford, UK: Oxford University Press; 2005.

[42] Rose MR, Charlesworth B. Genetics of life history in *Drosophila melanogaster*. II. Exploratory selection experiments. Genetics 1981;97:187–96.

[43] Rose MR, Vu LN, Park SU, Graves Jr JL. Selection on stress resistance increases longevity in *Drosophila melanogaster*. Exp Gerontol 1992;27:241–50.

8 3D Structure and Physiological Regulation of PAKs

Stefan Knapp

Nuffield Department of Clinical Medicine, University of Oxford, Old Road Campus Research Building, Structural Genomics Consortium, Oxford, UK

8.1 The PAK Family of Protein Kinases

p21-activated protein kinases (PAKs) are RAC/CDC42-dependent Ser/Thr kinases that play central roles in a wide range of cellular processes including regulation of proliferation, cell motility, morphology, and cytoskeletal dynamics and are expressed in a wide variety of tissues [1–4]. PAKs were originally identified as effectors for the GTPases, Cell division control protein 42 (CDC42), and RAC (Ras-related C3 botulinum toxin substrate), but a number of GTPase-independent pathways have also been reported that are regulated by this family of kinases. Based on their sequences and domain organization, the six PAKs in mammals have been grouped into two subfamilies: group I (PAK1–3) and group II (PAK4–6). Group I PAKs share a very high degree of sequence identity as well as many structural and functional similarities. Their N-terminal domain recruits regulators such as GTPases through GTPase-binding domain Cdc42/Rac Interactive Binding (CRIB), the non-catalytic region of Tyr kinase adaptor protein (NCK 1) adaptors through Pro-rich sequences and with PIX guanine exchange factors through an atypical Pro-rich sequence [5–7]. The highly conserved catalytic domain is located at the C-terminus. Group I PAKs are kept in an inactive state by autoinhibitory mechanisms that involve the N-terminal autoinhibitory domain (AID), which partly overlaps with the CRIB and inhibits PAK enzymatic activity by acting as a tightly binding pseudosubstrate. The AID also contains a dimerization domain, allowing autoinhibition in *trans* [8,9]. Group I PAKs also associate with an Src homology 3 (SH3) adaptor protein called PAK-interacting exchange factor (PIX). PIX is also associated with the scaffolding protein GIT1 (G-protein-coupled receptor kinase-interacting protein) and the cytoskeleton protein paxillin, targeting PAK to focal adhesions by a four-member signaling complex (PAK/PIX/GIT1/Paxillin), which has a well-established role in regulating focal adhesion turnover.

PAKs, RAC/CDC42 (p21)-activated Kinases. DOI: http://dx.doi.org/10.1016/B978-0-12-407198-8.00008-4

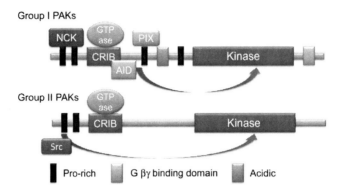

Figure 8.1 Schematic representation of regulatory domains in group I and group II PAKs and interaction partners. Interactions and regulatory functions of the highlighted domains are discussed in detail in the main text of this chapter.

The mechanisms that underlie the regulation of group II PAKs have been less clear since they contain no conserved autoregulatory switch domain [10]. However, group II enzymes also are regulated by N-terminal regulatory regions that inhibit catalytic activity. Group II PAKs contain GTPase-binding domains, but they are constitutively active even in the absence of a GTPase [1,11]. PAK4 binds preferentially not only to CDC42 but also to RAC, and interaction with these GTPases mainly has a targeting function. Coexpression of CDC42 results in translocation of PAK4 to the Golgi and induction of filopodia [1,12]. However, removal of the N-terminus results in an increase in kinase activity of PAK5, pointing to a regulatory function of the N-terminus of the kinase activity by intra- or intermolecular interactions for group II PAKs [13]. This hypothesis was recently confirmed by structural data that revealed an N-terminal autoinhibitory pseudosubstrate motif, which is discussed in more detail below [14]. A schematic representation of regulatory domains found in group I and group II PAKs is shown in Figure 8.1.

8.2   Structural Mechanisms of PAK Regulation

Kinases are highly dynamic molecules that are regulated by a large diversity of mechanisms. Experimental data on the structural changes that occur in kinases as a result of interactions with regulatory elements are essential for our understanding of kinase regulation by inter- and intramolecular mechanisms. Crystal structures of PAKs in their different activation states therefore make an essential contribution to our knowledge of the molecular mechanisms of PAK regulation. In addition, high-resolution structural data are essential for the rational design of potent and selective inhibitors.

All the protein kinases share the same overall architecture and an identical catalytic mechanism of adenosine triphosphate (ATP) γ-phosphate transfer. The mechanistic conservation of catalysis requires correct spatial arrangement within the catalytic domain of conserved catalytic motifs that are positioned in the kinase active site by a network of largely conserved interactions. Despite the large sequence diversity, the kinase active state is structurally highly conserved, allowing optimal alignment of motifs for high-affinity ATP binding and for the catalytic reaction.

The kinase fold consists of two domains that are also called lobes. The N-terminal lobe (N-lobe) usually comprises five β-sheets (β1–β5) and helix αC, which plays a key role in the regulation of kinase activity. The larger C-terminal lobe (C-lobe) mainly contains helical secondary structure and harbors the substrate binding site. Both kinase lobes are linked by a flexible hinge that anchors the cofactor ATP to the hinge backbone by a set of hydrogen bonds. ATP is additionally coordinated by a conserved and highly dynamic loop region, located between sheets β1 and β2, that contains a flexible Gly-rich sequence motif (GXGXφG, where φ is a Tyr or Phe residue), which is also called the phosphate binding loop or P-loop. The high flexibility of this loop region allows the P-loop to act like a gate that opens and closes the active site and tightly interacts with the ATP phosphates in the ATP-bound state. ATP is also coordinated by the tripeptide motif Asp–Phe–Gly (DFG), which interacts with the ATP phosphates via the Asp of DFG and a magnesium ion and is required for catalysis. A third interaction is important for high-affinity ATP binding: the conserved Lys residue located in the Val–Ile–Ala–Lys (VIAK) motif stabilizes the ATP phosphate groups by a salt bridge network that also involves a highly conserved Glu residue in helix αC. The salt bridge between the Lys of VIAK and the αC Glu is a hallmark of the kinase active state, as it is indicative that the regulatory αC helix is correctly positioned to allow high-affinity binding of ATP. Furthermore, in the active state of kinases, the Phe of DFG represents an anchor point that correctly positions the Asp of DFG and also tightly links the DFG motif with the αC helix that couples substrate recognition to cofactor binding [15].

Kinase substrates usually interact in an extended conformation with a shallow surface groove that is created by side chains located in the main C-terminal lobe helices (αD–αH) and in the C-terminal section of the activation segment (A-loop). The activation segment has a key role in kinase substrate recognition and in kinase activation. It usually comprises a 20–30 residue loop region, which is unstructured in kinases that are regulated by A-loop phosphorylation. The flexibility of the unphosphorylated A-loop results in partially unstructured substrate binding sites in inactive kinases, and at least partially impedes the substrate interactions. On activation by A-loop phosphorylation at specific sites, the activation segment assumes a well-ordered structure through formation of polar interaction with the catalytic loop tripeptide motif His-Arg-Asp (HRD) motif and other positively changed kinase residues. The main function of the activating phosphate moiety is therefore to neutralize the large positive charge around the substrate binding site. In contrast to the conserved and well-defined active state, a large diversity of conformations have been observed in inactive kinases, in which at least one of the key regulatory elements located in the kinase catalytic domain is usually displaced, unstructured, or sterically blocked [16].

Detailed structural and biochemical studies have elucidated the mechanism of PAK1 activation, defining a mechanism of regulation that is conserved in group I PAKs [17–21]. The crystal structure of inactive PAK1 reveals that residues of the kinase inhibitory switch domain (ISD) located N-terminal to the kinases domain bind to the

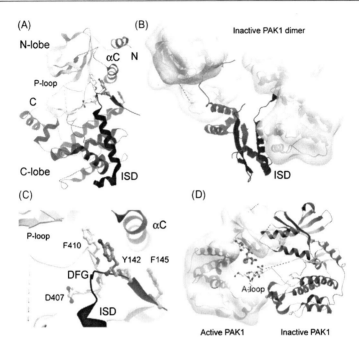

Figure 8.2 Conformations of PAK1. (A) Detailed view of autoinhibited PAK1 kinase. Shown is a ribbon diagram with the ISD binding (blue) to the substrate binding site. The main structural elements as well as the N- and C-termini are labeled. (B) Autoinhibited dimer. The two ISDs are colored blue and red and are shown as a ribbon diagram. The kinase domains are depicted in transparent surface representation. (C) Detailed view of the interaction of the ISD (blue) with the substrate binding site. The DFG motif is pushed into the active site, displacing helix αC. (D) Asymmetric PAK dimer in which one kinase domain is in an active conformation and the other is in an inactive conformation. The inactive kinase domain has an extended activation loop conformation that binds to the substrate binding site of the active kinase domain in the dimer in a conformation suitable for *trans* autophosphorylation of its activation segment. (For interpretation of the references to color in this figure legend, the reader is referred to the web version of this book.)

cleft between the two kinase lobes, acting as a pseudosubstrate [8]. Pseudosubstrate inhibition is a very common mechanism of kinase regulation and has been described for *cis* (in the same polypeptide chain) and *trans* (located in a regulatory protein) acting interactions. In autoinhibited PAK1, the ISD blocks substrate binding and stabilizes an inactive conformation by distorting regulatory elements in the kinase active site. In particular, interaction with the ISD distorts the activation segment, which is pushed deep into the ATP site, displacing αC, and disrupts the canonical salt bridge located between the Lys of VIAK and the conserved αC Glu (Figure 8.2).

In group I PAKs, the ISD partially overlaps with the GTPase-binding domain (CRIB) [2]. On kinase activation, the GTP-bound CDC42/RAC binds to the ISD region, releasing the pseudosubstrate from the substrate binding site, and triggers unfolding of the ISD region. Interestingly, the structure of autoinhibited PAK1 revealed an asymmetric dimer, but from the structural information alone, it was not clear whether the ISD of the same kinase molecule would inhibit the kinase domain in *cis*, or if the N-terminal ISD of the interacting kinase would act as a pseudosubstrate inhibitor in *trans*. Later on it was demonstrated that PAK1 also forms homodimers *in vivo* and that its dimerization is regulated by the intracellular level of GTP-CDC42 or GTP-RAC1. Dimeric PAK1 adopts a conformation in which the N-terminal inhibitory portion of one PAK1 molecule in the dimer inhibits the catalytic domain of the other in *trans*. Interestingly, only one GTPase molecule is sufficient for activation of both PAK1 kinases in the dimer [9]. Studies on PAK2 also showed that the phosphorylated kinase domain of PAK2 dimerizes in solution, and that this association can be prevented by addition of a substrate peptide. Mutagenesis data showed that dimerization of PAK2 is important for autophosphorylation at the activation loop [22].

Removal of the ISD by the GTPase releases the activation segment, allowing autophosphorylation at Thr423. Subsequently, the unphosphorylated PAK kinase domain forms an asymmetric dimer in which one protomer adopts an active conformation, acting as an activating kinase for the second molecule and providing evidence of the dynamic nature of protein kinases [23]. Once activated, phosphorylated PAK1 autophosphorylates several additional sites in the PAK1 N-terminus, preventing rebinding of the N-terminus to the catalytic domain and as a consequence autoinhibition of its catalytic activity [24].

The regulation of group II PAKs is less well defined. However, recent structural comparisons of monophosphorylated group II PAK crystal structures provide an insight into the plasticity of these kinases as well as the mechanism of the phosphotransfer and substrate-recognition processes [25]. In a unique mechanism, the comparison of different catalytic domain structures revealed rearrangements in helix αC that involve the generation of an additional N-terminal helical turn in this helix. The observed structural changes allow residues in helix αC to form interactions with different conserved residues in the kinase N-lobe, resulting in a dynamic linkage of the Gly-rich loop, helix αC, and the activation segment during the catalytic cycle. The observed movement also allows dynamic regulation of the αC helix, firmly anchoring it in an active conformation in the ATP-bound state. Thus, in group II PAKs the Gly-rich loop serves as a "molecular sensor" that detects ATP binding and consequently triggers structural rearrangements, resulting in closing of the kinase active site by a concerted movement of the Gly-rich loop—the hinge between the two kinase lobes—by N-terminal elongation of the αC helix. Together, these differences create distinct ATP and substrate binding pockets among the PAK isoforms. The N-terminal helical extension of αC is unique to group II PAKs because group I enzymes have a conserved Pro residue directly N-terminal to αC, which would not allow helical extension. The dynamic differences might be used for the development of group I and group II PAK-specific inhibitors. On a mechanistic level, the helical extension in group II enzymes results in a swinging movement of a conserved Arg residue (Arg359 in PAK4). When oriented toward the in N-terminus, this conserved arginine forms hydrogen bonds with the Gly-rich loop and stabilizes a closed conformation competent for ATP and ATP-mimetic inhibitor binding. On shortening of αC by one N-terminal turn, Arg359 is oriented toward the substrate binding site, forming hydrogen bonds with the phosphate moiety of the phosphorylated activation loop residue S474. In this conformation, a more open conformation of the Gly-rich loop is stabilized by a hydrogen bond between the αC residue Gln357 and the main chain oxygen of Thr332 in PAK5 [25].

Dynamic changes in the kinase catalytic domain are very different in group I PAKs. Comparison of crystal structures in the active and inactive conformations of the catalytic domain showed that the main structural rearrangement that facilitates the kinase domain to adopt a closed, active conformation is the "in" and "out" concerted swinging movement of αC helix with helix αA located N-terminal to the kinase domain. The outward movement of the helices provides an open conformation that allows binding of ATP, whereas the inward swinging results in a closed active conformation that is competent for catalysis.

Recent reports demonstrated that group II PAKs are also autoinhibited by interaction of the catalytic domain with an N-terminal autoinhibitory pseudosubstrate motif. Crystallographic and enzymatic data showed that full-length PAK4 is constitutively autoinhibited by a group II specific regulatory motif, suggesting a common type II PAK autoregulatory mechanism. In contrast to group I PAKs, autoinhibition is not released by GTPases but instead by interaction with SH3 domains such as the Src SH3 domain, which interacts with a Pro-rich sequence of the autoinhibitory group II domain [14]. The group II PAK AID is evolutionary highly conserved [26].

8.3 Alternative Mechanisms Leading to PAK Activation

A number of additional activation mechanisms that are independent of the canonical activation mechanism mediated by RAC/CDC42 have been described for group I PAKs. These mechanisms comprise activation by proteolytic cleavage of the PAK N-terminal ISD, phosphorylation of the activation segment by upstream kinases, interactions with SH3 domain-containing proteins and binding of sphingosine or several related long-chain sphingoid-based lipids [4,27,28].

Bokoch et al. [27]demonstrated that treatment of cells expressing wild-type PAK1 with sphingosine, fumonisin B, or sphingomyelinase induced increased activity of PAK1. Site-directed mutagenesis studies suggested that lipids act at a site that is overlapping or identical to the GTPase-binding domain on PAK1. In support of this notion, the inactive sphingosine derivative N,N-dimethylsphingosine has been shown to be an effective inhibitor of PAK1 activation in response to either sphingosine or CDC42 [27]. However, the process of PAK1 activation by lipids is probably more complex than a simple competitive displacement event of the ISD. Activation of cell surface receptor tyrosine kinases by insulin or growth factors results in activation of phosphatidylinositol 3-kinase (PI3K) and the generation of the second messenger, phosphatidylinositol 3,4,5-trisphosphate. This phospholipid messenger leads to the recruitment and activation of 3-phosphoinositide-dependent protein kinase-1 (PDK1) to the plasma membrane. It has been shown that PDK1 directly phosphorylates

PAK1 in the presence of sphingosine or other phospholipids at the regulatory activation segment residue Thr423, suggesting a phospholipid-dependent direct activation mechanism [29].

Group II PAKs are also activated via the PI3K signaling pathway. PAK4 activity is stimulated in response to hepatocyte growth factor (HGF), a multifunctional cytokine that binds and activates the Met/HGF receptor [30]. Signaling through the Met/HGF receptor plays an important role in chemotaxis, cell growth, and morphogenesis. Met receptor stimulation recruits PAK4 to the cell periphery but not specifically in lamellipodia, contributes to HGF-induced changes in cytoskeletal organization, and induces cell rounding [31]. Interestingly, since this process is mediated by group II PAKs, it does not require CDC42. However, while PI3K-specific inhibitors inhibit PAK4 activation and cell rounding, the isolated C-terminal kinase domain of PAK4 can induce cell rounding in the presence of specific PI3K inhibitors, suggesting that the N-terminal region acts as a negative regulator of PAK4 activity, which is in agreement with recent studies that identified a Pro-rich N-terminal inhibitory domain in PAK4 [14].

Studies on sphingosine-1-phosphate (S1P), and the G-protein-coupled receptor kinase 2 effector kinase showed that PAK activity can also be regulated by S1P through a GTPase-independent mechanism that results in stimulation of the RAC/PAK/MEK/ERK pathways [32]. Also, extracellular signal-regulated kinases 2 (ERK2) binding within the AID of PAK1 has been described in S1P-treated cells. ERK2 phosphorylates PAK1 on Thr212, leading to PAK1 activation [33].

Studies on PAK2 show that upon stimulation of cells by a variety of stresses, such as radiation- and chemical-induced DNA damage, hyperosmolarity, addition of sphingosine, serum starvation, and contact inhibition, PAK2 is proteolytically cleaved and activated by caspase 3 in a mechanism that dissociates the autoinhibitory N-terminus from the catalytic domain [34,35]. Caspase-induced activation of PAK2 has an important function in the regulation of apoptosis. Mice that lack the PAK2 caspase cleavage site, and can therefore not be activated by this protease, show a decreased cell death of primary mouse embryonic fibroblasts and an increased cell growth at high cell density. In addition, embryonic fibroblasts of these animals show reduced activation of the effector caspases 3, 6, and 7, indicating that caspase activation of PAK2 amplifies the apoptotic response through a positive feedback loop resulting in more activation of effector caspases [36]. In addition, PAK2 has been reported to bind to caspase 7, leading to inactivating phosphorylation at Ser-30, Thr-173, and Ser-239, protecting cells from apoptosis and contributing to chemotherapeutic resistance in human breast invasive ductal carcinoma [36].

Surprisingly, recent reports showed that PAK1 can be activated by PIX in a GTPase-independent manner. In this study, it was shown that the paxillin-associated adaptor GIT1 directly stimulates PAK autophosphorylation activity. Interaction with PAK requires the GIT N-terminal Arf-GAP domain, but surprisingly it does not require Arf-GTP activating protein (GAP) activity, suggesting that the GIT/PIX/PAK complex can function independent of GTPases [37]. For detailed information on PAK regulation (activation or inactivation) in living cells, see Chapters 3 and 4.

8.4   PAK Substrate Specificity and Downstream Signaling Events

Several studies reporting on PAK substrate specificity showed that both group I and group II PAKs have a preference for Arg residues at substrate $P - 1$ to $P - 5$ positions but that both PAK groups differ by selection of group-specific residues at $P + 2$ and $P + 3$ positions. ($P - n$ indicates the amino acid n residues N-terminal to the phosphorylated residue and $P + n$ indicates the amino acid n residues C-terminal to the phosphorylated residue.) [3,38].

Structural studies suggested that the N-terminal lobe loop located in the linker region, which assumes alpha helical structure during the group II PAK catalytic cycle as a result of the helical extension of helix αC, is important for group-specific substrate selectivity [4,8,25]. The group II-specific swinging movement of an arginine residue located in the αC helical extension region makes contact with bound substrate in one of the observed conformations. Mutagenesis data showed the importance of this PAK2-specific interaction: when PAK2 and PAK4 residues are swapped, the selectivity at $P + 2$ and $P + 3$ positions reverses [38]. However, given the dynamic nature and flexibility of the PAK kinase domains and potential effects of regulatory proteins, it is difficult to ascertain the stringent selectivity between the two groups in cellular systems. Indeed, subcellular localization, interaction partners, and surface accessibility are also important factors that will determine PAK substrate specificity (Figure 8.3).

One of the best-studied *in vivo* PAK substrates is LIM kinase-1 (LIMK1), which inhibits actin depolymerization by phosphorylating cofilin [39–41]. Actin filaments are key structures of cellular adhesive complexes, and the dynamic reorganization of actin filaments determines cell morphology. The actin depolymerizing factor (ADF) and cofilin are actin-binding proteins that regulate cytoskeletal dynamics by severing and depolymerizing actin filaments. ADF/cofilin is negatively regulated by phosphorylation by the ubiquitously expressed Ser/Thr kinase LIM domain kinase (LIMK), which is activated by phosphorylation on Thr508 by either the Rho kinase ROCK or the PAK [42]. Among the two isoforms of LIMK (LIMK1 and LIMK2), LIMK1 is mainly phosphorylated by PAK4. Moreover, dominant-negative LIMK1 inhibits most PAK4-mediated effects on the cytoskeleton that are induced by a constitutively active PAK4 mutants, suggesting that LIMK1 is the major downstream effector of PAK4 [12]. LIMK inactivates ADF/cofilin by phosphorylating cofilin at Ser3. Phosphorylation at this site recruits the scaffold protein 14-3-3ζ resulting in downregulation of the cofilin pathway [43]. The role of PAK in the activation of LIMK is additionally amplified by PAK-mediated inhibition of a phosphatase called slingshot (SSH1), which inactivates LIMK by dephosphorylation [44].

In addition to the formation of filopodia, another hallmark of PAK activity on the cytoskeleton is the dissolution of stress fibers. This process has been shown to be triggered through direct phosphorylation of myosin light-chain kinase (MLCK) by PAK1, resulting in decreased MLCK activity [45]. This process seems to be specific to PAK1. PAK4 does not phosphorylate MLCK. However, activation of PAK4 also results in stress fiber dissolution, suggesting that additional mechanisms exist. One of these pathways may involve GefH1, a Rho-specific guanine nucleotide exchange factor that is

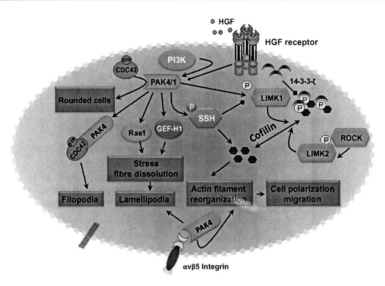

Figure 8.3 Multiple cellular substrates and activation mechanisms of PAK4. Kinases involved in the described pathways are highlighted in red and phenotypic changes are shown in boxes. (For interpretation of the references to color in this figure legend, the reader is referred to the web version of this book.)

phosphorylated by PAK4, inhibiting its nucleotide exchange activity and Rho-mediated signaling [46]. Another key substrate linked to cell mobility and adhesion is β-catenin, which is phosphorylated by both PAK1 and PAK4 [47,48], a process that may mediate anchorage-independent cell growth in PAK1/PAK4 overexpressing cancer cells. PAK4 has also been reported to phosphorylate the cytoplasmic tail of β-5 integrin on Ser-759 and Ser-762, two critical residues for integrin-mediated cell migration, giving further evidence of the key role of PAK kinases in regulating cell motility [49].

Cell growth and survival-associated functions of PAK kinases are mediated through stimulation of the Mitogen Activated Protein Kinase (MAPK) pathway as well as through phosphorylation of proteins mediating apoptotic responses, such as BCL2-associated agonist of cell death (BAD), which stimulates cell survival [50]. A key target of PAK kinases is the protein kinase Raf, which leads to its activation [51]. Raf-1 is regulated by phosphorylation of critical serine residues in the catalytic domain. PAK3 was shown to phosphorylate Raf-1 on Ser338, stimulating Raf-1 activation. For further discussion of PAK substrates, see Chapter 2.

Acknowledgments

The author is grateful for the financial support of the SGC, a registered charity (number 1097737) that receives funds from the Canadian Institutes for Health Research, the Canada Foundation for Innovation, Genome Canada, GlaxoSmithKline, Pfizer, Eli Lilly, Takeda, AbbVie, the Novartis Research Foundation, Boehringer Ingelheim, the Ontario Ministry of Research and Innovation, and the Wellcome Trust.

References

[1] Abo A, Qu J, Cammarano MS, Dan C, Fritsch A, Baud V, et al. PAK4, a novel effector for Cdc42Hs, is implicated in the reorganization of the actin cytoskeleton and in the formation of filopodia. EMBO J 1998;17(22):6527–40.

[2] Bokoch GM. Biology of the p21-activated kinases. Ann Rev Biochem 2003;72:743–81.

[3] Kumar R, Gururaj AE, Barnes CJ. p21-activated kinases in cancer. Nat Rev Cancer 2006;6(6):459–71.

[4] Eswaran J, Soundararajan M, Kumar R, Knapp S. UnPAKing the class differences among p21-activated kinases. Trends Biochem Sci 2008;33(8):394–403.

[5] Burbelo PD, Drechsel D, Hall A. A conserved binding motif defines numerous candidate target proteins for both Cdc42 and Rac GTPases. J Biol Chem 1995;270(49):29071–29074.

[6] Bagrodia S, Taylor SJ, Jordon KA, Van Aelst L, Cerione RA. A novel regulator of p21-activated kinases. J Biol Chem 1998;273(37):23633–23636.

[7] Thevenot E, Moreau AW, Rousseau V, Combeau G, Domenichini F, Jacquet C, et al. p21-activated kinase 3 (PAK3) protein regulates synaptic transmission through its interaction with the Nck2/Grb4 protein adaptor. J Biol Chem 2011;286(46):40044–40059.

[8] Lei M, Lu W, Meng W, Parrini MC, Eck MJ, Mayer BJ, et al. Structure of PAK1 in an autoinhibited conformation reveals a multistage activation switch. Cell 2000;102(3):387–97.

[9] Parrini MC, Lei M, Harrison SC, Mayer BJ. Pak1 kinase homodimers are autoinhibited in trans and dissociated upon activation by Cdc42 and Rac1. Mol Cell 2002;9(1):73–83.

[10] Jaffer ZM, Chernoff J. p21-activated kinases: three more join the Pak. Int J Biochem Cell Biol 2002;34(7):713–7.

[11] Cotteret S, Jaffer ZM, Beeser A, Chernoff J. p21-activated kinase 5 (Pak5) localizes to mitochondria and inhibits apoptosis by phosphorylating BAD. Mol Cell Biol 2003;23(16):5526–39.

[12] Dan C, Kelly A, Bernard O, Minden A. Cytoskeletal changes regulated by the PAK4 serine/threonine kinase are mediated by LIM kinase 1 and cofilin. J Biol Chem 2001;276(34):32115–32121.

[13] Ching YP, Leong VY, Wong CM, Kung HF. Identification of an autoinhibitory domain of p21-activated protein kinase 5. J Biol Chem 2003;278(36):33621–33624.

[14] Ha BH, Davis MJ, Chen C, Lou HJ, Gao J, Zhang R, et al. Type II p21-activated kinases (PAKs) are regulated by an autoinhibitory pseudosubstrate. Proc Nat Acad Sci USA 2012;109(40):16107–12.

[15] Yamaguchi H, Hendrickson WA. Structural basis for activation of human lymphocyte kinase Lck upon tyrosine phosphorylation. Nature 1996;384(6608):484–9.

[16] Huse M, Kuriyan J. The conformational plasticity of protein kinases. Cell 2002;109(3):275–82.

[17] Gizachew D, Guo W, Chohan KK, Sutcliffe MJ, Oswald RE. Structure of the complex of Cdc42Hs with a peptide derived from P-21 activated kinase. Biochemistry 2000;39(14):3963–71.

[18] Leeuw T, Wu C, Schrag JD, Whiteway M, Thomas DY, Leberer E. Interaction of a G-protein beta-subunit with a conserved sequence in Ste20/PAK family protein kinases. Nature 1998;391(6663):191–5.

[19] Morreale A, Venkatesan M, Mott HR, Owen D, Nietlispach D, Lowe PN, et al. Structure of Cdc42 bound to the GTPase binding domain of PAK. Nat Struct Biol 2000;7(5):384–8.

[20] Thompson G, Chalk PA, Lowe PN. Interaction of PAK with Rac: determination of a minimum binding domain on PAK. Biochem Soc Trans 1997;25(3):509S.

[21] Thompson G, Owen D, Chalk PA, Lowe PN. Delineation of the Cdc42/Rac-binding domain of p21-activated kinase. Biochemistry 1998;37(21):7885–91.

[22] Pirruccello M, Sondermann H, Pelton JG, Pellicena P, Hoelz A, Chernoff J, et al. A dimeric kinase assembly underlying autophosphorylation in the p21 activated kinases. J Mol Biol 2006;361(2):312–26.

[23] Wang J, Wu JW, Wang ZX. Structural insights into the autoactivation mechanism of p21-activated protein kinase. Structure 2011;19(12):1752–61.

[24] Chong C, Tan L, Lim L, Manser E. The mechanism of PAK activation. Autophosphorylation events in both regulatory and kinase domains control activity. J Biol Chem 2001;276(20):17347–17353.

[25] Eswaran J, Lee WH, Debreczeni JE, Filippakopoulos P, Turnbull A, Fedorov O, et al. Crystal structures of the p21-activated kinases PAK4, PAK5, and PAK6 reveal catalytic domain plasticity of active group II PAKs. Structure 2007;15(2):201–13.

[26] Baskaran Y, Ng YW, Selamat W, Ling FT, Manser E. Group I and II mammalian PAKs have different modes of activation by Cdc42. EMBO Rep 2012;13(7):653–9.

[27] Bokoch GM, Reilly AM, Daniels RH, King CC, Olivera A, Spiegel S, et al. A GTPase-independent mechanism of p21-activated kinase activation. Regulation by sphingosine and other biologically active lipids. J Biol Chem 1998;273(14):8137–44.

[28] Roig J, Tuazon PT, Traugh JA. Cdc42-independent activation and translocation of the cytostatic p21-activated protein kinase gamma-PAK by sphingosine. FEBS Lett 2001;507(2):195–9.

[29] King CC, Gardiner EM, Zenke FT, Bohl BP, Newton AC, Hemmings BA, et al. p21-activated kinase (PAK1) is phosphorylated and activated by 3-phosphoinositide-dependent kinase-1 (PDK1). J Biol Chem 2000;275(52):41201–41209.

[30] Benvenuti S, Comoglio PM. The MET receptor tyrosine kinase in invasion and metastasis. J Cell Phys 2007;213(2):316–25.

[31] Wells CM, Abo A, Ridley AJ. PAK4 is activated via PI3K in HGF-stimulated epithelial cells. J Cell Sci 2002;115(Pt 20):3947–56.

[32] Penela P, Ribas C, Aymerich I, Eijkelkamp N, Barreiro O, Heijnen CJ, et al. G protein-coupled receptor kinase 2 positively regulates epithelial cell migration. EMBO J 2008;27(8):1206–18.

[33] Sundberg-Smith LJ, Doherty JT, Mack CP, Taylor JM. Adhesion stimulates direct PAK1/ERK2 association and leads to ERK-dependent PAK1 Thr212 phosphorylation. J Biol Chem 2005;280(3):2055–64.

[34] Roig J, Traugh JA. Cytostatic p21 G protein-activated protein kinase gamma-PAK. Vitam Horm 2001;62:167–98.

[35] Walter BN, Huang Z, Jakobi R, Tuazon PT, Alnemri ES, Litwack G, et al. Cleavage and activation of p21-activated protein kinase gamma-PAK by CPP32 (caspase 3). Effects of autophosphorylation on activity. J Biol Chem 1998;273(44):28733–28739.

[36] Marlin JW, Chang YW, Ober M, Handy A, Xu W, Jakobi R. Functional PAK-2 knockout and replacement with a caspase cleavage-deficient mutant in mice reveals differential requirements of full-length PAK-2 and caspase-activated PAK-2p34. Mamm Genome 2011;22(5–6):306–17.

[37] Loo TH, Ng YW, Lim L, Manser E. GIT1 activates p21-activated kinase through a mechanism independent of p21 binding. Mol Cell Biol 2004;24(9):3849–59.

[38] Rennefahrt UE, Deacon SW, Parker SA, Devarajan K, Beeser A, Chernoff J, et al. Specificity profiling of Pak kinases allows identification of novel phosphorylation sites. J Biol Chem 2007;282(21):15667–15678.

[39] Soosairajah J, Maiti S, Wiggan O, Sarmiere P, Moussi N, Sarcevic B, et al. Interplay between components of a novel LIM kinase-slingshot phosphatase complex regulates cofilin. EMBO J 2005;24(3):473–86.

[40] Arber S, Barbayannis FA, Hanser H, Schneider C, Stanyon CA, Bernard O, et al. Regulation of actin dynamics through phosphorylation of cofilin by LIM-kinase. Nature 1998;393(6687):805–9.

[41] Kuhn TB, Meberg PJ, Brown MD, Bernstein BW, Minamide LS, Jensen JR, et al. Regulating actin dynamics in neuronal growth cones by ADF/cofilin and rho family GTPases. J Neurobiol 2000;44(2):126–44.

[42] Edwards DC, Sanders LC, Bokoch GM, Gill GN. Activation of LIM-kinase by Pak1 couples Rac/Cdc42 GTPase signalling to actin cytoskeletal dynamics. Nat Cell Biol 1999;1(5):253–9.

[43] Gohla A, Bokoch GM. 14-3-3 regulates actin dynamics by stabilizing phosphorylated cofilin. Curr Biol 2002;12(19):1704–10.

[44] Niwa R, Nagata-Ohashi K, Takeichi M, Mizuno K, Uemura T. Control of actin reorganization by slingshot, a family of phosphatases that dephosphorylate ADF/cofilin. Cell 2002;108(2):233–46.

[45] Sanders LC, Matsumura F, Bokoch GM, de Lanerolle P. Inhibition of myosin light chain kinase by p21-activated kinase. Science 1999;283(5410):2083–5.

[46] Callow MG, Zozulya S, Gishizky ML, Jallal B, Smeal T. PAK4 mediates morphological changes through the regulation of GEF-H1. J Cell Sci 2005;118(Pt 9):1861–72.

[47] Wong LE, Reynolds AB, Dissanayaka NT, Minden A. p120-catenin is a binding partner and substrate for group B Pak kinases. J Cell Biochem 2010;110(5):1244–54.

[48] Dohn MR, Brown MV, Reynolds AB. An essential role for p120-catenin in Src- and Rac1-mediated anchorage-independent cell growth. J Cell Biol 2009;184(3):437–50.

[49] Li Z, Zhang H, Lundin L, Thullberg M, Liu Y, Wang Y, et al. p21-activated kinase 4 phosphorylation of integrin beta5 Ser-759 and Ser-762 regulates cell migration. J Biol Chem 2010;285(31):23699–23710.

[50] Schurmann A, Mooney AF, Sanders LC, Sells MA, Wang HG, Reed JC, et al. p21-activated kinase 1 phosphorylates the death agonist bad and protects cells from apoptosis. Mol Cell Biol 2000;20(2):453–61.

[51] King AJ, Sun H, Diaz B, Barnard D, Miao W, Bagrodia S, et al. The protein kinase Pak3 positively regulates Raf-1 activity through phosphorylation of serine 338. Nature 1998;396(6707):180–3.

Epilogue: "Lateral Thinking" Is the Key to a Great Leap in Biomedical Sciences

Hiroshi Maruta

NF/TSC Cure Organisation, Melbourne, Australia

On October 8, 2012, the Nobel Committee at Karolinska Institute in Stockholm announced that Drs. Shinya Nakayama (50) at Kyoto University and John Gurdon (79) at Cambridge University (Figure 1) would share the 2012 Nobel Prize in physiology/medicine for their pioneer work in the field of stem cell research. http://www.nobelprize.org

Back in 1962, John Gurdon, then a Ph.D. student at Oxford University, challenged a bold experiment using mature cells from adult frogs in an attempt to prove his hypothesis that nuclei in any mature cells of the same body contain the exactly same set of genes as the nuclei in the original egg. He replaced the nuclei in eggs

Figure 1 The 2012 Nobel laureates in physiology/medicine. John Gurdon (left) and Shinya Nakayama (right).

Figure 2 Charles Darwin (1809–1882).

with those from mature intestinal cells of adult frogs, and successfully created healthy adult frogs from these transgenic eggs, clearly indicating that any mature nucleated cells could be reprogrammed into multipotential stem cells. His groundbreaking 1962 paper (Dev, Biol 4, 256–73) states, "Adult frogs derived from nuclei of single somatic cells."

In other words, nucleated cells from different organs contain the same set of genes, including the *PAK*1 gene. So if abnormally activated PAK1 causes cancer in the pancreas or colon, it could potentially cause cancer in any other organ. That is exactly what we, molecular oncologists, currently see in human and other mammals. Traditionally, the majority of clinical oncologists used to see pancreatic cancer quite differently from colon or breast cancers, and tried to treat cancers in distinct organs differently. However, as discussed in Chapters 2 and 3, all solid tumors in a variety of organs depend on PAK1 for their growth, and therefore a set of PAK1 blockers could cure a variety of cancers of entirely distinct organs.

Finding such a common cause of cancers in distinct organs is a very general biological approach called "lateral thinking," which was initiated by Charles Darwin (Figure 2) to explore the origin of species back in 1859. If I understand correctly, almost every gene in the human body shares 99% of its sequence homology with its counterpart in the chimpanzee, clearly proving Darwin's postulation that these two primates share a common ancestor. Accordingly, around 1974, Sydney Brenner (Figure 3), then the director of the MRC Molecular Biology Laboratory at the University of Cambridge, UK, started using a tiny worm (nematode) called *Caenorhabditis elegans* as a new experimental animal. This worm consists of only around 1000 cells, including mainly muscle, neuronal, reproductive, and digestive cells. His original dream was to map the entire neuronal cell network, because this

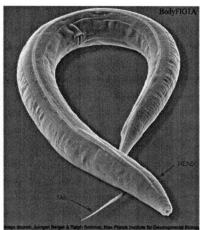

Figure 3 Sydney Brenner and *C. elegans.*

worm is transparent. Although his dream has not come true as yet, his *C. elegans* genetic research brought a series of very fruitful results, and he shared the 2002 Nobel Prize in physiology/medicine with two of his colleagues for their pioneer work on this worm. If I understand correctly, like the human genome, the whole genome of this worm has been sequenced, and interestingly this worm also contains a *PAK1* gene that is functionally almost indistinguishable from the human *PAK1* gene.

We recently found a few interesting phenotypes of a PAK1-deficient mutant of this worm called RB689, including reduced fertility (smaller litter size), higher heat tolerance, and longer lifespan, as discussed in Chapter 7. In other words, using these PAK1-deficient phenotypes we are able to screen for a variety of effective PAK1 blockers or elixirs among synthetic chemicals and natural products in a way that is both very quick and less expensive than mouse models. Among natural products, we identified a few PAK1 blockers such as curcumin and salidroside as well as caffeic acid phenethyl ester (CAPE) and artepillin C (ARC). The latter two are the major anticancer ingredients in propolis, an inexpensive bee product. Furthermore, these natural PAK1 blockers are known to suppress depression (make us feel happy) and suppress a few other neuronal disorders, as discussed in Chapter 5. So it is not a big surprise to guess why, for 100 million years, bees have been so addicted to using the natural products of specific plants, such as poplar trees, to make honeycombs (Figure 4) to protect their larvae.

Back in 1975, our old friends at NIH, Bob Adelstein and his colleague Mary Anne Conti [1] (Figure 5) found that human platelet myosin II (double-headed) requires the phosphorylation of its regulatory light chain by a myosin light-chain kinase (MLCK) for F-actin-activation of its intrinsic $Mg^{2+}ATPase$ activity. At that time our own team at NIH was working on an unconventional myosin called myosin I (single-headed) from a soil amoeba called *Acanthamoeba castellanii* whose

Figure 4 Bees and honeycombs.

intrinsic Mg^{2+}ATPase activity is not activated by F-actin without an extra protein called "cofactor protein". Making a direct analogy to platelet myosin II, we "laterally" thought that the cofactor might be an MLCK. However, it did not phosphorylate the light chain of the amoeba myosin I. Two years later, we were so surprised to discover that the amoeba cofactor did phosphorylate the myosin I, but at its heavy chain! [2]. To our further surprise, around two decades later this amoeba myosin I heavy-chain kinase (MIHCK) was found to be the very first member of the PAK family [3,4], establishing that PAKs play a key role in mediating the RAC/ CDC42-induced dynamics of actomyosin-based cytoskeleton in all eukaryotes, from amoeba/ yeast to human.

Lastly, I would like to present the latest example of how "lateral thinking" was used to solve a mystery concerning a multipotent antidepressant called desipramine (see Figure 6). This year a group led by Yongsoon Kim [5] at the Nevada Cancer Institute in Las Vegas discovered that a few mutants of *C. elegans* that lack *ASM* (acid sphingomyelinase) 1–3 genes can live longer than their control counterparts. ASM is a lysosomal enzyme that hydrolyzes sphingomyelin to produce ceramide, a sphingolipid that eventually activates PAK1–3. As discussed in Chapter 7, a PAK1-deficient strain of *C. elegans* (RB689) also lives longer than the control, clearly indicating that PAK1 shortens the lifespan of this worm by inactivating the longevity protein FOXO. Interestingly, desipramine inhibits ASM, eventually inactivating PAK1–3. Thus, there should be no surprise in their latest finding: desipramine also extends the lifespan of this worm in a FOXO-dependent manner [5]. As discussed in Chapters 5–7, several antidepressants such as CAPE and curcumin can suppress the growth of cancers, inflammation, and Alzheimer's disease as well as extend the lifespan of this worm by blocking PAK1. In other words, blocking either ASM or PAK1 will lead to basically the same pharmacological effects and is a step toward the improvement of our quality of life (QOL).

Figure 5 Bob Adelstein's Lab. Bob (seated row, second from right) and Mary Anne (standing row, third from left). Kindly provided by Dr. Bob Adelstein.

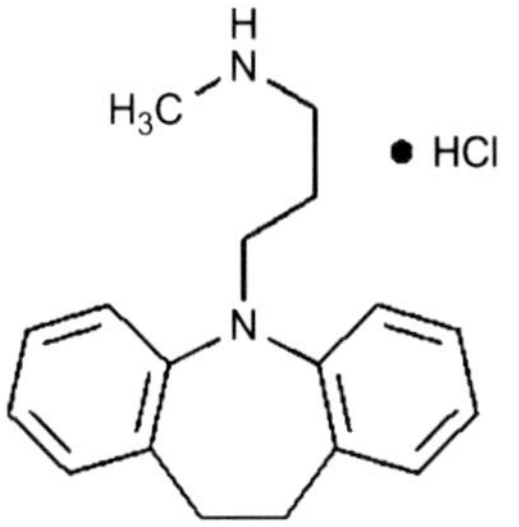

Figure 6 Chemical structure of desipramine.

In conclusion, based on such "lateral thinking," I strongly believe these PAK1 blockers could help us build a disease-free world, substantially reduce the world population, survive global warming, and enjoy a longer lifespan with a higher QOL, and above all feel far happier.

References

[1] Adelstein RS, Conti MA. Phosphorylation of platelet myosin increases actin-activated ATPase activity. Nature 1975;256:597–8.

[2] Maruta H, Korn E. *Acanthamoeba* cofactor protein is a heavy chain kinase required for actin activation of the Mg^{2+}-ATPase activity of *Acanthamoeba* myosin I. J Biol Chem 1977;252:8329–32.

[3] Lee SF, Egelhof T, Mahasneh A, Cote G. Cloning and characterization of a *Dictyostelium* myosin I heavy chain kinase activated by CDC42 and RAC. J Biol Chem 1996;271:27044–27048.

[4] Brzeska H, Szczepanowska J, Hoey J, Korn E. The catalytic domain of *Acanthamoeba* MIHCK. II. Expression of active catalytic domain and sequence homology to PAK. J Biol Chem 1996;271:27056–27062.

[5] Kim Y, Sun H. ASM-3 acid sphingomyelinase functions as a positive regulator of the DAF-2/AGE-1 signaling pathway and serves as a novel Anti-Aging target. PLoS One 2012;7:e45890.

About the Authors

Greg M. Cole

Professor Greg. M. Cole received a B.Sc. in physics/biochemistry in 1973, and a Ph.D. in physiology from the University of California, Berkeley, CA, in 1986.

He was a postdoctoral fellow at UCSD (1986–1989), and eventually joined the faculty at UCLA in 1994. In 1996, in collaboration with Dr. Karen Hsiao-Ashe's group at the University of Minnesota, his team helped develop the first successful academic transgenic mouse model for Alzheimer's disease (AD). Based in part on screens from Drs. Cole and Frautschy in preclinical models, four NSAID-related compounds—ibuprofen, R-flurbiprofen, curcumin, and docosahexaenoic acid (DHA)—have advanced to clinical trials for Alzheimer's. Because of NSAID side effects, his team investigated safer alternatives including DHA, an omega-3 fatty acid derived from fish, and identified DHA's role in preventing amyloid formation and alpha–beta toxicity to excitatory synapses involving PAK. They have also explored the efficacy of the turmeric compound curcumin to control inflammation, oxidative damage and the cellular stress response and act on soluble toxic abeta and tau species. They developed a bioavailable formulation of curcumin that is now in clinical trials for Alzheimer's and other diseases. Their primary goal is to develop tools for the prevention of Alzheimer's and possibly other degenerative diseases of aging. He is currently full professor in the Department of Medicine and Neurology and the Alzheimer's Disease Research Center at UCLA.

Graham Côté

Professor Graham Côté received a Ph.D. in biochemistry from the University of Alberta in Edmonton, AB, Canada, under the supervision of Dr. Larry Smillie in 1980. His career highlights include postdoctoral studies (1980–1984) at Dr. Ed Korn's lab at the NIH, where he studied the role of heavy-chain phosphorylation in the regulation of *Acanthamoeba* myosin II. In 1984, he was appointed an assistant professor in the Department of Biochemistry at Queen's University in Kingston, ON, Canada and in 1996 was promoted to full professor.

His expertise is focused on the regulation of myosin activity by protein kinases. His team discovered the Dictyostelium myosin II heavy-chain kinase (MHCK-A), which phosphorylates the tail of myosin II, inhibiting filament assembly. MHCK-A is the founding member of the atypical alpha-kinase family, which have a novel type of catalytic domain distinct from conventional protein kinases. He also found that Dictyostelium class I myosins are activated by members of the PAK family and has identified several novel small light chains that bind to the myosin I neck region.

Sally A. Frautschy
Professor Sally A. Frautschy received an M.Sc. in physiology from the University of California at Davis, CA, in 1980 and a Ph.D. for her work on chronic stress from the University of Guelph, ON, Canada, in 1986.

She completed two postdoctoral fellowships, at the University of California at San Diego (UCSD) and at the Scripps Institute in La Jolla, CA, where she received RO1 funding. She moved to a faculty position at the University of California at Los Angeles (UCLA) in 1994. She has served on NIH and Veteran's Administration study sections since 1996. She is associate editor of the *Journal of Neuroinflammation* and won the Alzheimer's Association Los Angeles award in 1999. She has presented at national and international conferences and has published over 86 articles in peer-reviewed papers. She was one of the initial pioneers in showing an inflammatory response to amyloid in Alzheimer's models and to develop treatments for this disease prior to transgenic models. She is also one of the first investigators to use these models to develop drug interventions (notably bioavailable curcumin) and to understand mechanisms of the disease. She is currently based at the Alzheimer's Research Lab at the Department of Medicine and Neurology at UCLA.

Hong He
In 1994, Hong He received a Ph.D. from the University of Melbourne, Parkville, Melbourne, Australia, and in 1987 she received an M.D. from Beijing Medical University (now called Health and Science Center of Peking University), Beijing, China. From 1994 to 2003 she was an SRS at the Tumor Suppressor Lab of the Ludwig Institute for Cancer Research in Melbourne, Australia. Since 2003, she has been an SRS in the Cancer Biology Lab at the Department of Surgery, University of Melbourne, Austin Health, Melbourne. Areas of expertise include molecular biology of oncogenic RAS-RAC/CDC42-PAK1-beta catenin signaling pathways and development of a series of PAK1 blockers such as AG 879/GL-2003 and PP1, which are potentially useful for therapy of RAS cancers such as pancreatic, colon, and stomach cancers.

Ramesh K. Jha
Ramesh Jha received an M.Tech. in Biochemical Engineering and Biotechnology from the Indian Institute of Technology, Delhi, India in 2002, and a Ph.D. in computational design of novel proteins for affinity reagents from the University of North Carolina at Chapel Hill, NC, in 2010.

In 2002, he held the position of research assistant at AstraZeneca R&D, Bangalore, India, researching assay development for high-throughput screening. In 2010, he became a postdoctoral research associate at Los Alamos National Laboratory, Los Alamos, NM. His research interests include protein engineering for novel and improved functions with applications in bioenergy and biosecurity and the structural aspects of PAKs.

Stefan Knapp
Professor Stefan Knapp studied chemistry at the University of Marburg, Germany and the University of Illinois, Urbana, IL. He received a Ph.D. in protein

crystallography at the Karolinska Institute in Stockholm, Sweden (1996) and continued his career at the Karolinska Institute as a postdoctoral scientist (1996–1999). In 1999, he joined the Pharmacia Corporation as a principal research scientist in structural biology and biophysics and left the company in 2004 after the acquisition of Pharmacia by Pfizer to set up a research group at the Structural Genomics Consortium in Oxford, UK. Since 2008, he is a professor at the Nuffield Department of Clinical Medicine at Oxford University, UK, and since 2012, he is the director for chemical biology at the Target Discovery Institute. His research interests are structural mechanisms that regulate signal transduction pathways, in particular protein kinases such as PAKs, and the rational design and development of selective inhibitors that target these kinases. A particular focus of his research team is also the development of protein interaction inhibitors that target epigenetic reader domains such as bromodomains.

Qiulan Ma
Qiulan Ma is currently an SRS at the Alzheimer's Research Laboratory in the Department of Neurology at UCLA. She received an M.D. in 1987 and an M.Sc. in 1996 from Harbin Medical University, China. In 1999, she received a Ph.D. in molecular genetics of Alzheimer's and Parkinson's diseases from Sun Yat-Sen University of Medical Sciences, China.

In 1999, she joined a postdoctoral training program at Dr. Piu Chan's lab at the Beijing Institute of Geriatrics, Department of Neurology at Xuanwu Hospital, part of the Capital University of Medical Sciences in Beijing. This is China's biggest research center for both clinical and basic research on AD. After 2 years of postdoctoral training, she was offered a tenured associate professor position in the Department of Neurology of Xuanwu Hospital. In 2000, she joined Dr. Kenji Uéda's laboratory at the Tokyo Institute of Psychiatry in Japan. In 2004, she joined Dr. Greg Cole's lab at UCLA, and since then has focused on Aβ oligomer-induced PAK1-LIMK signaling defects in AD pathogenesis, mechanisms of DHA/curcumin intervention and beta-amyloid clearance and passive immunization therapeutics in AD.

Hiroshi Maruta
Professor Hiroshi Maruta received his Ph.D. from the Department of Pharmaceutical Sciences, University of Tokyo Graduate School, Japan, in 1972. He is currently manager of the NF/TSC Cure Organization, a nonprofit in Melbourne, Australia.

His career highlights include working as a National Institutes of Health (NIH) international research fellow, Department of Molecular, Cellular and Developmental Biology at the University of Colorado, Boulder, CO; a visiting associate at the NIH at Bethesda, MD (1974–1980); and a senior research scientist (SRS) at the Max Planck Institute of Biochemistry and Psychiatry in Martinsried, Germany. In 1988, he became head of the Tumor Suppressor Laboratory at the Ludwig Institute for Cancer Research in Melbourne, Australia, and since 2006 he has been a visiting professor at Hamburg University Hospital in Eppendorf, Germany and the University of Maryland at Baltimore, MD. His areas of expertise include biochemistry of actomyosin-based cytoskeleton, molecular biology of oncogenic RAS/PAK signaling pathways, and development of PAK1 blockers that are potentially useful for clinical application.

Shanta M. Messerli

Dr. Messerli received a B.Sc. in Neuroscience from Wellesley College, Wellesley, MA, in 1994 and a Ph.D. from Purdue University, West Lafayette, IN, in the area of neuronal injury and regeneration in 2001.

In 2004, she completed postdoctoral training at Massachusetts General Hospital, Harvard Medical School in the development of a gene therapy for neurofibromatosis type 2 (NF2) using mouse models, and in 2007, she completed another postdoctoral training at the Marine Biological Laboratory (MBL), Woods Hole, MA in the area of mechanisms of drug resistance. In 2008, she became assistant research scientist at the Cellular Dynamics Program at MBL.

Her principal contribution to tumor biology was to prove *in vivo*, using mouse models of NF2 and glioma, that a series of natural PAK1 blockers such as propolis (CAPE-based Bio 30 from New Zealand and ARC-based green propolis extract from Brazil) and antimalarial drugs suppress almost completely the PAK1-dependent growth of these brain tumors.

Masato Okada

Professor Masato Okada studied at the Institute for Protein Research, Osaka University, Japan, receiving a Ph.D. in 1988 for his work on Tyr kinase in neuronal cells. Prior to this, he received a B.Sc. in Biochemistry from the Faculty of Science of Kyoto University, Japan, in 1985.

Career highlights include the position of assistant professor at the Division of Protein Metabolism at the Institute for Protein Research, Osaka University, Japan, and appointment as a visiting scientist at Jon Cooper's lab at the Fred Hutchinson Cancer Research Center in Seattle, WA. In 1996, he became associate professor at the Division of Protein Metabolism, Institute for Protein Research, Osaka University, and in 2000 he was promoted to full professor at the Department of Oncogene Research of the Research Institute for Microbial Diseases at Osaka University. His areas of expertise include molecular biology of Tyr kinases such as SRC family and CSK (C-terminal SRC kinase) as well as PAKs.

Sumino Yanase

Associate professor Sumino Yanase graduated from the Chemistry Department, Faculty of Science, Tokai University, Japan, in 1987 and the Graduate School of Integrated Science, Yokohama City University, Japan, in 1995. She received a Ph.D. from Tokai University School of Medicine in 1999. In 2000, she worked as a postdoctoral fellow at Tom Johnson's lab at the University of Colorado at Boulder.

She has held positions at Toranomon Hospital, Tokyo, Japan, Kanagawa Prefectural College of Nursing and Medical Technology and the Kanagawa Prefectural College of Public Health in Japan. In 2005, she joined the faculty of Daito Bunka University School of Sports and Health Science in Japan. She started working on *C. elegans* in 1995, when she was a Ph.D. student in Dr. Naoaki Ishii's laboratory at Tokai University in Japan.

Fusheng Yang

Fusheng Yang is currently a staff research associate IV in the Department of Neurology at UCLA. He received an M.D. in 1983 and an M.Sc. in 1989 from Hunan Medical University, China. He joined UCSD in 1993 as a research fellow in Dr. Greg Cole's lab, and after a few months moved to UCLA with Dr. Cole. Since then, he has played a central role in animal intervention studies with NSAIDs, omega-3 fatty acids, curcumin, and other disease-modifying interventions including PAK, BACE, and tau aggregation inhibitors for AD pathology.

CPSIA information can be obtained at www.ICGtesting.com
Printed in the USA
BVOW01*0616310713

327116BV00006B/84/P